n a Pumpe in a Tubbe.

ALSO BY THALASSA CRUSO

Making Things Grow: A Practical Guide for the Indoor Gardener

Making Things Grow Outdoors

These are Borzoi Books,
published in New York
by Alfred A. Knopf

To Everything There Is A Season

To Everything There Is A Season

The Gardening Year with Thalassa Cruso

Alfred A. Knopf
New York
1973

THIS IS A BORZOI BOOK
PUBLISHED BY ALFRED A. KNOPF, INC.

Published in the United States by Alfred A. Knopf, Inc., New York,
and simultaneously in Canada by Random House of Canada Limited,
Toronto. Distributed by Random House, Inc., New York.

Library of Congress Cataloging in Publication Data

Cruso, Thalassa. To everything there is a season.

1. Gardening. I. Title.
SB455.C74 1973 635.9 73–5959
ISBN 0–394–48074–0

Manufactured in the United States of America

First Edition

To the three sets of three sisters
who in different generations
have loved and worked with plants

Julia, Caroline, and Mary-Theresa
Janet, Mildred, and Amian
and now
Ala, Sophia, and Thalassa

Preface

These essays originally appeared in a much more condensed form in the Boston Sunday *Globe* and in *McCall's* Magazine, and I am very grateful to both these publications for permission to reprint. They were written over a period of five years and cover a variety of overlapping springs, summers, falls, and winters and reflect the many aspects of gardening in New England.

Writing a regular gardening column is very different from writing a gardening book. Gardening books are not concerned with *if*s and *and*s, or complaints about the weather; they speak with a calm authority that transcends all such matters. But a horticultural column, unless it is written entirely along the lines of what to do when, can be much more flexible in content and more revealing about the personality of the author. For it describes, week after week, what actually is going on as well as what ought to be happening. Even when at a later date the immediacy of a situation as outlined in a column is frozen into a factual essay, the sense of urgency and improvisation remains. A garden column can and should take into account the vagaries of the weather and how they are affecting conventional garden practice and advice. And since a column must appear on schedule whether the writer has any immediate pertinent horticultural advice to offer or not, it often ranges quite far afield into areas related to horticulture but not specifically concerned with gardens or potted plants. For in a gardening column, and in the essays into which the columns may eventually be converted, what become clear are the peripheral activities and personality traits at work turning the writer into a successful gardener. The reader discovers the writer's likes and dislikes, strengths and weaknesses. The announcement of a success brings pleasure; a failure is a disappointment shared—all because of the continuing association that builds up a sense of friendship between the writer and the reader. And although the writer may be a perfect stranger to you in person, you feel you have learned a lot about him or her through such columns.

I found this out many years ago when I was a dedicated reader of the gardening column written by the late Vita Sackville-West in the London Sunday *Observer*. Vita Sackville-West was a distinguished, prolific writer on many subjects as well as a superb horticulturist who could convey her enthusiasm in print: her columns set a standard beyond the reach of most of us. Her free-ranging concern with the living world as a whole makes just as delightful reading today as when it was written—and I still learn something every time I reread her books. Nothing would give me more pleasure than to feel that my essays share some of these same qualities. For they too are eclectic in the sense that although they circle plants and gardens, they do not confine themselves only to specifics. In them I cover a multitude of matters related to the art of growing things as events take place—which is not necessarily how I expect those events to take place! The columns mirror the everyday day-to-day triumphs, curiosities, frustrations, and pleasures that afflict every gardener who tries to be in tune with the world and who also attempts to strike a balance between nature and what appears to be its most ancient and determined adversary—man.

We speak of playing golf or tennis but of working in the garden, but that's not quite true in my case. I have never played at anything, be it tennis, T.V., or archaeology; I have always worked at it, for such is my nature. But the word "work" is a misnomer for any subject that engrosses me: work is not a dreary matter of unwilling labor, but wholehearted involvement that takes the rough with the smooth and teaches a skill.

I am still an enormous distance from wholly mastering the art of growing things and the more difficult process of looking around me with a truly observant eye. But these essays show how I have tried both to sharpen my perceptions and to increase my horticultural abilities, and I hope they also show what fun it has all been. For horticulture, in its innumerable manifestations, is a hobby that can go with you anywhere at any time, just like your shadow. No one interested in plants and the living world can ever be bored.

Rewriting and reorganizing your own work is a tedious process: a column structured to fit into a specific compressed space often balks when it has to be changed into a careful essay. And since every writer has personal favorites among his work, the process of elimination is also a painful one. In these conflicts, no one could have been better or kinder than my editor, Jane Garrett. She helped me with the winnowing, spurred me on with the rewriting, and listened patiently to my protestations. She also put me finally and forever in her debt by tracking down the delightful illustrations, for which we are deeply grateful to The Boston Athenaeum for *The Gardeners Labyrinth* (London, 1608), and the Metropolitan Museum of Art, Harris Brisbane Dick Fund, 1945, for *Le grant kalendrier et compost des Bergiers* (Troyes, 1529), and to Charles V. Passela of the Morgan Library for his superb photography of the illustrations and to Carole Lowenstein for her skillful use of them.

Behind every book there are always other unseen hands. My agent, Ellen Neuwald, provided steady, unflagging enthusiasm when it was merited and tactful criticism when that was needed. The care and nurture of writers is her trade, and she is unfailingly brilliant at it. Peter Hotton, garden editor of the Boston *Globe*, must also receive his due. The editing he performs on the original columns, which is not always related to the limitations of space, has often shown me a better way to make a point. Mary Anne O'Roark of *McCall's* Magazine must also be thanked for many kindnesses. My husband bore patiently with me when I took a pile of unedited material on a so-called relaxing holiday, and my many colleagues at Station WGBH, Boston, went out of their way to bend their schedules to my deadlines. Without all this help I would never have finished.

This is the third time I have written on horticultural matters. Each book has been a challenge and each has called for a different approach. The freedom to write at will on any aspect of horticulture, which led to these essays, is a freedom I particularly cherish and enjoy. If something of this feeling comes through to the reader of this book I will be greatly content.

Thalassa Cruso
Boston 1973

Contents

Preface

January

The Muddled Seasons *2*
Waiting *5*
Paper-Whites *7*
The Verdict *10*
They Always Know *11*
Ideas *14*
Trees *16*

February

Philodendrons *20*
The Woods *24*
Snowdrops *25*
Clivias *29*
Toads *34*
Birds *36*
The Vine *38*

March

John 42
Salt? 44
Records 46
New Ideas 49
The Enclosure 52
Change 54
Tomatoes 56
Alice and Forsythia 59

April

Potted Bulbs 64
Roads 67
The Dilemma 71
Easter Lilies 74
Ivies 77
The Pond 80
Heaths and Heathers 85
The Brides 89

May

The Rites of Spring 94
Continuity 96

Tulip Time 98
Easy and Attractive 101
Nettles 104
Trials and Tribulations 107
Try Some Orchid Cacti 112

June

Summer House Plants 118
Regal Geraniums 123
Where Have All the Flowers Gone . . . 128
Petunias and Ivy 132
Winter Geraniums 136
I Remember, I Remember . . . 140
The Intruders 143

July

Biennials 148
The Salt Marsh 153
Water 155
Naturally 159
Vacation Time 161
Let's Not Spoil It Again 164
Loosestrife 168

August

Summer Wild Flowers *172*
Not with Me! *176*
My Word! *180*
The Three Sisters *184*
Stripes *190*
Tuberous Begonias *193*

September

Gloxinias *202*
Wildlife *209*
Swan Song *213*
Country Gardens *220*
Fall Duties *225*

October

Regeneration *230*
Chrysanthemums *234*
The Extravaganza *240*
South African Bulbs *242*
Don't Stop Yet *248*

November

Conservation at Home *256*
Memories *260*
Amaryllis *262*
Individuality *267*
Exasperation *271*
A Parrot *275*

December

Hollies *280*
Waiting *284*
For Those with Eyes to See *285*
Associations *288*
Christmas *292*
Waste Not *295*
Enchantment *298*

January

The Muddled Seasons

As a child I loved fairy stories, and like most little girls, I used occasionally to dream about falling into an enchanted sleep and remaining suspended in time until the inevitable prince arrived. And had any such magic process been available in my youth, when awakened I would have been able to tell instantly the season of the year from the fresh flowers that loving hands—I was a sentimental child—had kept constantly around my sleeping form. If I had been roused in deep winter there would have been tightly bundled Parma violets scenting the air; in very early spring primroses and the St. Brigid bright-colored anemones would have replaced them. Late spring would have brought the light fragrance of lilies of the valley and daffodils, and the high color of tulips. During the summer there would have been huge bouquets of roses and sweet peas, and as the year began to turn, carnations, lilies, dahlias, and gladiolus would have taken their place. The rest of the year chrysanthemums would have carried the floral burden. Flowers in my childhood told one, without equivocation, the season; they blossomed on a timetable that was linked to the movements of the sun and the heat of the air, a schedule that had ruled their bud set from the period of their most remote ancestors.

But if my granddaughters ever indulge in the same fantasies, they are going to be far more confused about the time of year when they wake up. Today many flowers are no longer tied to their ancestral flowering season; the skills of modern scientists have changed all that. Through the manipulation of light hours, combined with special cooling techniques, we can now have roses, carnations, sweet peas, gladiolus, and chrysanthemums all year round, and the list of plants whose flowering patterns can be changed is constantly expanding. Today home gardeners can also order daffodils, lilies, and lilies of the valley treated with a special cooling process that enables them to be timed to flower at some exact date that may be months removed from their normal blossom time.

In many ways this is all very delightful. It gives us a far greater choice of flowers than used to be available during the dreary winter months except at impossibly high prices, but there are also some disadvantages. Some of the potted plants that will set bud and come into bloom out of season under specialized treatment either collapse entirely and forever when they come into the house or else, removed from the artificial forcing program, revert to their original life style as soon as the premature flower is finished and refuse to produce any more buds until what to them is the natural flowering season comes around.

I always long to warn potential buyers not to take home those heavily forced small azaleas that appear on the market with many buds and a few unhappy flowers in time for Christmas or the New Year. These have undergone such intensive unnatural treatment to produce those early flowers that they will either refuse to open their remaining buds once they get into our houses or drop their leaves in a depressing shower. Often they combine both unhappy processes, and after a short unsatisfactory period indoors there is no solution but to throw them away, which is a great waste of money. The same problem occurs with gloxinias that are sold full of bloom and bud at this time of year. They have been brought into this condition through intensive and complicated forcing that involves a special combination of artificial light, humidity, and warmth which cannot be reproduced in anyone's house or greenhouse. When you get these gloxinias home, no matter how loving and learned your care, either they will droop miserably and never revive or all those promising buds will brown and rot off.

Primroses that appear on the market from November onward with one small unhappy bloom on a miserable little plant are an example of overforced plants that may not die in your house but will stubbornly refuse to provide you with any more flowers until their natural period of bloom rolls around, which is around mid-February under glass in New England. And by the time the primrose is ready to put on a spectacular show, you may have become so tired of its squat, do-nothing, dumpy appearance that you have already thrown it out!

African violets belong also in this second category. These are plants that need a certain amount of strong sunless light to flower. Unless you can provide it by artificial lighting, there will be no more flowers on fall-blossoming African violets, which were specially processed by those who raise them to produce this bloom, until the natural light in a north or east window is strong enough and long enough to induce bud set.

As a good general rule, buy flowering potted plants as near to their natural blossoming season as possible, and the simplest way to judge this is to buy flowering plants only when the stores are full of flourishing specimens. With this simple precaution you will get far more for your money than the temporary lift given your spirits by an out-of-season rarity.

Waiting

There's a stimulating feeling in the cold cellar at this time of year in spite of its deplorable appearance, for there is such a sense of latent energy imprisoned in that room. We rest a great many plants in this cool, dimly lit place; some go in even before we come back finally from the country. The first to be incarcerated are the Christmas cacti, which go into a dark, waterless resting period around the first of October. This plant will flower for me at the holiday season only if it is given a month with twelve hours of completely uninterrupted darkness every single night without even a streetlight nearby. Without such special handling my Christmas cacti flower either much later in the year or not at all. In mid-October, the clivias, hydrangeas, and geraniums go into the same room together with all the summer-flowering bulbous plants. Achimenes, tuberous begonias, and the gladiolus corms sit quietly in withered disarray on the shelves giving no trouble, but the gloxinias which are also stashed away in that same cool cellar have to be watched, so that they can be rewatered and set under artificial lights as soon as new growth shows. You have to watch carefully, for the growth first appears as minute fuzzy pink spots. And if you don't start to water gloxinias as soon as this tentative growth shows, the corm may become discouraged and relapse into suspended animation from which it is almost impossible to revive it.

Hydrangeas also have to have an eye kept on them, for these are tricky plants to rest. They need to stay outdoors until the first light frosts kill the leaves, but they must not be left outdoors in bitter cold, which will kill the canes on which the fat buds for next year's growth already exist. Once indoors and leafless, hydrangeas need a period of rest, but if they are left entirely waterless the canes will die, and if watering is too profuse the plants will leaf out prematurely and set no flowers. To try to hit the happy medium, at which I am by no means always successful, I give the hydrangeas a trickle of water about once every ten days until March. After that date they come into the greenhouse for regular forcing and are repotted after they have leafed out. The geraniums look frightful; the old leaves are all dead and the pot soil bone dry. Yet there are plenty of signs of life in spite of all this misery, for although the stout stems have begun to shrink a little as the reserve of water stored in them is drawn upon, each is studded with pallid little shoots longing for the increase of light and water that will send them leaping into growth for the summer show.

The clivias are sitting around in a state of ill-concealed impatience. These plants do not really need dormancy, for their arched green foliage does not wither off during an enforced rest. They can in fact stay unharmed in the warm house, but in that case they will flower when and if they want. I happen to like flowers from the clivias in March to fill the plant windows, and I have found that I can achieve controlled blossoming by keeping the plants in the half-dark with only a trickle of water until after the New Year and then transferring them to a poorly lighted position under the benches in a very cold greenhouse. But by now they are bursting to make that change—there is light green growth at the base of the inner leaves which is the sign of onrushing vitality.

To us, winter at its worst has only just begun, but not so for the plants. They sense that the shortest day has already come and gone and the new growing season has started, and they are itching to get going.

Paper-Whites

A few years ago I tried some simple experiments with paper-white bulbs usually listed as polyanthus narcissus, although technically they should be known as N. Tazetta. They are some of the easiest bulbs to force indoors, and are consequently often much mishandled. The first, most obvious advice is to avoid, as you would the plague, any trick prepackaged arrangement in which the bulb is already set into some sort of container with instructions that the grower has only to add water and, like instant coffee, the results will be superb. This is just not so, and a great many people have been much disappointed with these offerings, for the bulbs are usually wretchedly inadequate both in size and in quality. If they can be induced to flower at all, the blossoms will be poor and the flower stem miserably weak. It cannot be stressed often enough that shrunken debilitated bulbs will never produce fine flowers and could not do so even if they were planted out in the celestial fields! The only certain way to have fine flowers from bulbs is to buy the very best quality from an established dealer, and to get them as early in the season as possible so that you are not left with the pickings that wiser shoppers have rejected.

In my small tests I was anxious to discover whether there would be any marked difference in the quality of the flowers, the strength of their leaves and stems, and the timing of the opening to full flowering, when good bulbs, all bought from the same source, were grown in identical containers under exactly the same conditions but using different growing mediums. Since most paper-white bulbs are grown in solid containers without any drainage vents, I tried mine the same way. I happened to have a large number of identical vases, left over from a daughter's wedding, which were oblong in shape and about six inches deep. The different rooting mediums consisted of stones and water, damp perlite, damp vermiculite, and imported bulb fiber (available in the East from White Flower Farm in Litchfield, Connecticut). I also used my own rich compost, and a container filled with store-bought sterile soil. In each container the bulbs were planted in what seems to be the accepted pattern in this country of burying the base of the bulb just below the surface of the soil and leaving the rest exposed. For interest's sake I also included a container with a three-inch layer of roofing pebbles to form a dry well at the bottom, with an inch of rough half-made compost over that; this I planted English fashion, setting the bulbs on the layer of compost, and piling more loosely around them almost to their necks.

Since I was interested only in the quality of the flowers and the strength of the leaves and stems and not how the containers did in the house after planting, they all went straight into the cool greenhouse close to the glass. Considering that the flower bud is already fully formed in embryo inside a bulb when it is bought and that all my test examples were carefully chosen for identical size and good condition, the variations in their subsequent behavior were fascinating. Those in the pebbles and water and in the damp perlite and vermiculite followed that aggravating course of rising on stilts formed by their own roots. I had always supposed that this tendency came from growing the bulbs in inadequate light, but in a greenhouse this could not be the factor. Perhaps it is connected with lack of nourishment in the growing medium. Grown in strong light, these stiltlike specimens held the flowers well above the foliage, but the individual bulbs in each container opened unevenly: some were almost finished blooming as others were just unfurling. The stems were strong, but the leaves flopped badly and those particular mediums made staking impossible. The overall appearance was unattractive, with the bulbs twisted about lopsidedly on their elevated root systems. Keeping the water at the correct level among the pebbles was a nuisance: too little damaged the flowers; too much opened up the possibility of rot. Getting the proper balance of moisture was exceptionally hard with both perlite and vermiculite. These volcanic materials hold water like a sponge and must in consequence be watched carefully and continuously. I would not recommend any of these growing methods as suitable for the casual gardener. The bulbs in undrained containers with compost or sterile soil for the growing medium did very poorly. Again it was impossible to prevent the soil from getting too wet, and in the container with compost the roots rotted. The sterile soil, being very fine, alternated; it either got too wet or else dried out unexpectedly quickly on warm days in the greenhouse, and it was particularly tiresome to manage.

Of all the artificial mediums, the imported bulb fiber, which contains fertilizer as part of its composition, proved far the most successful. The bulbs planted in it started off more slowly than the others, and they flowered last. But the roots never rose above the surface of the soil, and the fiber stayed evenly moist—I was able to leave it without attention for several days at a time, which was not the case with any of the other test specimens. The flowers opened evenly; the leaves and stems were strong and well colored and, as they elongated, proved easy to stake. It seems a great pity that this material is not more freely available in this country, for I have grown hyacinths and daffodils equally successfully in it. However, all is not lost for those who cannot chase down imported material; for far the best results came from the container with the layer of pebbles underneath the compost in which the bulbs were buried deeply. The roots stayed safely under the compost, and although the topsoil dried out a little in the heat of the greenhouse, there was evidently enough storage water among those pebbles to provide adequate nourishment for the roots but not enough to cause rot. The bulbs all flowered at the same time and at the same height without need of staking, and looked like an advertisement for paper-whites, an effect I rarely achieve.

I thought this all might have been a series of happy coincidences, something that would never occur again, but in subsequent years I have planted my paper-whites in this same manner and achieved the same spectacular results as long as there was a sufficiently deep layer of stones to catch the surplus water in the undrained container.

The Verdict

The past summer, following as it did upon a deplorable winter, was not an easy one for gardeners. Not only was there massive snow damage to be repaired, but the rain and winds that plagued us through July and August made it hard to conceal any outstanding deficiencies among the perennial plants brought about by those same unending snows, for the annuals that I bought to fill in various gaping holes simply would not grow. The physical problems were therefore considerable in themselves, and emotionally they were greatly compounded by an impending visit from my brother, an accomplished gardener who lives in England, for whom I always try to put on a good show—which in this past season amounted to an impossibility.

We like to inspect each other's horticultural efforts at fairly regular intervals, and since we are a competitive family, these visitations are taken pretty seriously. Not that we ever comment adversely on anything the other does; that would not be cricket—criticism when needed is offered either by silence or by suggesting an alternative method of growing some particular plant.

In spite of my deplorable garden I was looking forward to the visit, for I was anxious to get my brother's reaction to some new landscaping ideas we had installed since he was last over. I also wanted to discover, without actually asking, whether he had managed to flower a rather temperamental clematis we both grow which, providentially, was putting on a tremendous show for me.

The week before his arrival a roaring gale blew away a great deal of the bloom from this one high spot of my garden, but enough blossom remained for me to lead my brother casually near to where it was flowering and stand around expectantly. I received an oblique answer to my unasked question with the comment that he wondered whether that clematis wouldn't do even better if its roots were more shaded—which leads me to conclude that his has done far better than mine!

The main test, however, was the new perennial and annual bed. This we had recently reorganized into a formal pattern of curved lines for easier management, and I was slightly uneasy about the possible reaction to this change. Perennial borders with their billowing free-form interlocking groups of plants are the glory of English gardens, and though the new, highly disciplined design was entirely valid for our purpose, I still felt a little as though I had let the team down by such a radical revision.

Nothing much was said at the time, other than some polite praise for the variety of plants I had crammed into the space. The battered condition of the annuals received the silence that was their due. But months later I got a letter with my brother's projected planting plan for a big new border he was laying out, and to my considerable amusement I saw it too was to be designed with curved lines exactly like mine.

Naturally there was no comment on this innovation, only a request for information about the use of certain plants, but I must admit to a slight sense of triumph in seeing my ideas calmly appropriated. Unlike Lucy and Linus, I am the younger of the pair, and it is not often that I emerge the clear winner!

They Always Know

I frequently read of experiments designed to illustrate the amount of sensory perception possessed by plants in our love/

hate relationship with them. This is a very hard matter to pin down in scientific terms, but those of us who work with plants don't really need learned papers to prove to us that they can be both obstinate and perverse.

We all understand, for example, that delicate hybrids are far harder to raise than their sturdier ancestors—this is the price we pay for better plants. You and I no doubt would find it extremely hard to live under the conditions that were commonplace in winter even for our grandfathers; the chances are that we would be miserably uncomfortable from a combination of constant cold and indigestion. We realize that we seem to have turned into more delicate human beings than our forebears and need rather different handling. But plants refuse to be as consistent as we are; otherwise, how could delicate hybrids, which perished miserably last winter on heated cables under carefully adjusted artificial lights, have turned up in the pot soil of some of last year's cyclamen? The intruders, which are columneas, were never deliberately planted in those pots, and for the life of me I can't imagine how they got there. And now that they are completely inextractible, some of them seem to be setting buds under conditions of cold and light which contradict everything these plants are supposed to require.

Ferns in abundance regularly spring up among the stems of my Thanksgiving cacti. This is another piece of improvident behavior on the part of a rather picky plant. Following my habit with all tree cacti, I keep the pots rather dry between waterings, and this is the very last life style that ferns are supposed to demand. Furthermore, I cannot imagine how they have managed to evolve in those pots. Ferns turn into recognizable plants only after a long and rather complicated embryonic stage during which they don't look like anything at all attractive or worth saving. I have had those old cacti for many years, and I don't repot them very often. I do, however, keep the surface of the pot soil clean and top-dressed, and I cannot imagine how several varieties of fern species I have never possessed escape that regular weeding. Even more irritatingly, they have managed to do so for several years, so that I now have Thanksgiving cacti that look like old-fashioned poinsettias with a ruff of ferns around their legs!

I have a friend who grows the most difficult of the indoor ferns, the maidenhair, to perfection. She often gives a party when they are at their superb best for us all to admire the extravagant lacy foliage. I would dearly love equal maidenhair elegance in my plant windows, and I tried for several years to grow them through very careful planting experiments with their spores. But after three years of complete failure I gave up in despair. Now suddenly an elderly gloxinia in the greenhouse has yielded no fewer than a dozen small maidenhair ferns from under its leaves in the last year. I can only assume that some of those experimental spores were washed or wafted into the pot, but even so I cannot imagine how these fussy little plants survived the vicissitudes of living under the leaves of a withering plant that was going through its annual drying-out period. There are three different leaf types among these infant plants, and I am growing them all individually with great pleasure. But I do feel a little like an unsuspecting sparrow that found three cuckoos in her nest!

This year's irritation is an orchid kindly left at my house by a TV viewer who raises these plants indoors and wanted me to try one too. She brought a lovely warmth-loving flowering plant with full instructions about managing it which I followed faithfully for several years without in truth getting much response out of that plant. This fall when we returned from the shore I gave up the struggle and put the orchid in the cold greenhouse under conditions quite unlike those its previous owner had suggested. Not long ago I got a letter from the donor asking how her orchid had done, and rather embarrassedly I wrote back and confessed the truth: not only had it failed to flower when handled according to her instructions, but I now had given up the attempt and taken the plant into a position to which I well realized it was not suited.

Hardly had that letter been mailed than the orchid threw a fat bud, the first it had ever produced under my care, and it has now been in flower for several weeks with a number of fine blossoms. All it needed to set it on its way was for me to admit defeat!

Ideas

One of the exciting things about gardening is the amount of unconscious knowledge that is assimilated almost without our knowing that we are learning. In my own case, a great deal of my basic horticultural knowledge is an accumulation of what I saw done in my youth, what might be called traditional lore, combined with what I have learned since through experiments and reading. And happily this educational process not only never ends but is also extremely stimulating. A new idea should make you want to go out and try it for yourself, for if and when you cease to be an adventurous gardener, you also cease to be an active gardener in the best sense.

Recently, for example, I discovered how to induce a cutting that had been rooted in water to make a successful transfer to living in soil, something I have always found rather difficult. For one thing, any water-rooted cutting gets hopelessly entangled into a thin stringlike muddle of intertwined roots as soon as it is taken out of the water. And the attempt to untangle the roots using even the finest-pointed pencil usually ends either by injuring the roots or by being such an utter failure that the miserable mess goes into the soil as is—and not unnaturally the cutting then fails to thrive. The solution is so obvious that I can't imagine why no one told me of it before, and because I have not lost a water-rooted cutting since I started to use this method, I want to pass it on.

The trick is this. Once any cutting has thrown out a good number of strong roots you should always try to wean it off the water diet. If you do not, the root hairs will modify or change so that they can extract needed oxygen from the water. Once this has happened, even if you can manage to move them into soil without too much damage, the shock of trying to extract the essential oxygen from air pockets instead of water often is fatal. The foolproof way to solve both the moving problem and the need to accustom the root hairs slowly to changing from water to air is to start by shifting the rooted cutting from a necked bottle in which it probably was rooted, for it is easier to support the plant that way, into something with a much wider mouth—one of those plastic old-fashioned glasses will do excellently. Fill the glass up with water, put a couple of pencils, or two small stakes, parallel to each other across the top of the glass, and hang the cutting, suspended by the leaves, between the pencils in the center of the glass with the roots happily spread out in water. Next, cautiously over a couple of weeks, slowly displace the water in the glass with fine soil; the kind that you can buy prepackaged is fine. The glass had better stand on a saucer during this period, for the displaced water will, of course, slop out. Once the glass is completely filled with soil that slowly has engulfed the roots, the pencils can eventually be removed, for the cutting should soon be able to stand alone. Leave it in the glass for a couple more weeks to allow the sodden soil to dry out. The drying process enables the root hairs to make the successful transfer from water to soil, but don't allow the soil to get bone dry during this process or you will ruin the experiment. You can tell when the cutting is happily reestablished by the appearance of new roots which can be seen through the transparent walls of the container. If you are getting impatient, try giving the cutting a slight tug. If it resists, you can be fairly certain that it has anchored itself into the earth with new roots. At that stage, what is now a successfully transplanted little plant can be knocked out of the plastic glass and potted into something more suitable, with proper drainage, and without any shock to the root system. I have found this method so easy and successful that I cannot resist suggesting that all windowsill gardeners try it. I wish that I could give the author of the admirable scheme proper credit for a brilliant idea.

But even as I do this let me offer a warning. Plants do not behave the same way for everybody—goodness knows I have found this out with ivies, which everyone can grow successfully indoors except myself! If your first experience with this method is not a total success don't give up—try again. Very few gardening triumphs come the first time around; it's persistence that counts.

Trees

During the last visit to the country garden it was too cold and blustery for me to work outdoors, so I did a lot of standing at the windows trying to decide how successful we had been in setting out evergreen shrubs to give winter interest to the grounds as a whole and particularly as seen from the house.

In general the plans have worked out well. The various junipers, yews, chamaecyparis, and arborvitae all make good strong punctuation marks against the leafless hedges as well as protecting the yard from the bitter winds off the sea. When the last few specimens go in next fall—they will mainly consist of low-spreading varieties to face down the tall specimens—a long-term project will at last be over.

The evergreens were chosen partly for winter interest, partly to tie the design together (something that they do extremely well), and partly to help provide extra food and cover for the birds during the winter. We cannot put out a feeder in that yard since we are there so irregularly, and I want to encourage birds to winter over because of the useful function they perform in keeping down insect pests. The new evergreens seem to have succeeded admirably in this latter purpose, for the birds are far more numerous and diverse than in the past—though the enthusiasm with which I saw innumerable birds working among the tree branches in our little woodlot suggests that somehow we have also encouraged a bumper crop of insects to winter over as well!

But what struck me even more forcefully on this occasion was the magnificence of the bare silhouettes of the branches of the big old trees against the wintry sky. I had never before properly appreciated how distinctive they are from one another, not only in shape but in the winter color of the bark. The elm, which worries us a great deal since it is now one of the few thriving specimens still alive in the village, grows in the traditional wineglass shape with a tall, bare trunk that is greenish gray. The copper beech that has grown into what might be called a stupendous tree sweeps its branches right to the ground, and the bark is smooth and light gray. Down by the waterfront are two tall lindens. These show black and dark when they are bare of leaves and grow in a narrow shape that in body language would say "don't touch," and they make a delightful contrast to the spreading heads of the other trees nearby.

The kousa dogwoods near the house are thick with pointed brown buds. This Oriental relative of our own dogwood blooms almost three weeks to the day after the petals of the local variety have fallen. And although the flowers are not unlike our own, the tree itself grows in a thicker, more closely knit fashion, while the bark, which is light gray and almost as smooth as a beech, peels away in patches showing a lighter color underneath that gives the tree great distinction. I recently read a distinguished English horticultural writer who had the highest possible praise for the kousa dogwood—and I fervently associate myself with his sentiments. However, unexpectedly he warned against ever pruning this particular variety of dogwood, maintaining that "the slightest touch of the knife will kill it." Perhaps that happens in England, but it certainly is not the case here. I regularly prune my kousa dogwoods in order to expose the delightful peeling main trunk, and if there are any aftereffects from this treatment, I think it stimulates bloom rather than killing the tree. My old kousas flowered better than ever after a rather unfortunate incident in which their roots were mistakenly root-pruned. And this past summer I cut all the low branches off one of the offspring of these old trees which, though it had grown quite large, had never favored me with any flowers. This winter, though it may be nothing but coincidence, the little tree is loaded with flower buds for the first time in its short life.

Last summer all the big trees were heavily pruned to provide less resistance to wind. At the time I thought the tree men were rather overdoing it, for the immediate aftereffect was extremely scanty. But now the opened-up trees are a delight: cleared of dead wood and overlapping branches, they look clean and healthy and free of encumbrances—a delight to a housebound gardener.

February

Philodendrons

Rather to my surprise I find myself being accused in print of not appreciating philodendrons sufficiently—which goes to show the unfairness of life, for in my time I have also been rebuked on TV for talking too much about them!

Philodendron is the family name for an enormous group of heat-loving plants which do well indoors in spite of our dry house air. They originate mainly in the tropical rain forests of the New World, where they can be seen scrambling up trees with stupendous vitality to reach the light above the jungle canopy. In the wild they carry rather uninteresting calla-lily-like flowers in a poison green color and trail thick ropes of aerial roots. These hanging roots sprout from the stems of the plant as it climbs upward, and they lengthen until they touch the ground, where they root and send extra nourishment back to the plant.

The largest of these climbing philodendrons, which is (alas) the type most often seen reproduced as an artificial plant, is the large split-leaf variety that some of us know as a monstera or Swiss cheese plant. There also exist hybrid forms of this large-leafed plant called "self-heading," a term that means that they no longer climb—that tendency has been bred out of them—but instead expand slowly sideways. In this diversified family there are many other varieties, some with solid spear-shaped leaves, others with reddish stems, some with foliage so heavily split that it is very lacy, and some with variegated leaves. The more unusual varieties take some searching out and can be expensive when tracked down, but they are all excellent plants for the indoor gardener, for they retain their urbanity under extremely trying conditions. One very common complaint about the large split-leaf variety is that the new leaves as they unfurl are solid without the interesting air holes. This can sometimes happen with the new leaves put out by a young plant, but more often it comes from insufficient light. An easy way to solve this difficulty, if you don't want to move the plant, is to provide it with extra hours of artificial light, perhaps by means of a small spotlight. This will produce a very dramatic effect as well as improving the health of the exhibit.

I grow a great many philodendrons with loving appreciation; my only difficulty is that they repay my affection by trying to drive me out of the house. To deal with their overexuberance I cut off all those untidy aerial roots which look very scruffy indoors, and I make up for the fact that I have deprived the plant of the special nourishment the roots would send back by giving an occasional dose of liquid fertilizer when new leaves are unfurling. I also cut off branches which are climbing over the tops of the frames and use these as cuttings to make new plants. Nothing could be simpler than producing a new philodendron: cut a length of stem with some aerial roots still attached and stuff the severed stem and the roots into a small pot with good internal drainage and plenty of rich soil. One of the problems is getting a heavy branch in a small pot to stand upright—the trick is to put the pot into a larger pot and fill the interstices between the two pots with pebbles. With careful watering and regular light missing, you will have a thriving new plant in a few weeks.

One of the advantages of philodendrons is that they can be kept in quite small pots in spite of their strong top growth; there is no need for constant repotting. If the old soil looks dreadfully depleted, remove the top three inches and add a fresh layer: this treatment will keep the plant happy for years. The main difficulty with a large-leaf vining monstera is getting it a support to climb upon that will fit into a small pot. I find those bark-covered slabs sold at enormous cost by florists extremely hard to balance in a small pot; it is really easier to put in two stakes—something about the length and dimension of broomsticks does very well. These should slant outward away from each other, and a length of chicken wire can be worked down each pole for a back rest. Use wire clippers to trim off the excess wire at the narrow converging ends of the support, and fold the clipped ends of the wire around each support so there will be nothing on which you can impale yourself. Indoors the wire will not heat up and hurt the plant, and the shoots can be threaded through it while they are still supple for a very pleasant effect. If you want to use a big trellis in a large pot, which is the way I grow two of my oldest Swiss cheese plants, use big stones to brace the back of the trellis inside the pot; otherwise it will never hold steady. If you have a self-heading philodendron that is growing much too big this can be separated. Sometimes the offsets can be pulled easily apart with no damage; more often division will have to be done with a knife. But think twice before you do this job on an expensive plant, for it has been my experience that these hybrids take an enormous time to recover from surgery.

To most people, the term philodendron means that small trailing plant with thin heart-shaped leaves that is available everywhere from the dime store upward. This is usually deplorably grown and is, I am afraid, so casually mistreated because it is inexpensive and easily replaced. Trailing little philodendrons with stringy stems and a few battered leaves are some of the most depressing and familiar of all horticultural sights, and it's a shame to mistreat a useful plant this way. Don't let it drip down; it too is a climber and will respond with enthusiasm if it is allowed to go upward. You can get a spectacular effect by tying string around a window frame and allowing the little plant to twine its way up. Treated this way as a substitute curtain it will turn into a delightful window arrangement while simultaneously presenting you with an insoluble problem about ever taking a holiday! If that seems too much trouble, try winding the trailing stems round and round on top of the pot soil and pinning them into the earth with hairpins. The green crown that will follow this treatment will be a revelation.

There is perhaps one word of warning that should be given about philodendrons: their leaves are poisonous to people and animals. As a matter of common sense, all children as soon as they reach a teachable age should be made to understand that they must never put any leaf in their mouths. With tiny children around, keep all plants well out of their reach—and no matter how sloppy your housekeeping may be in every other respect, always pick up every dead leaf from the floor! There is also a problem with cats, which get desperate for something green to eat if they are never let outside. Philodendrons will poison cats slowly but surely if they start to eat the leaves. Cats, however, prefer sprengeri, the so-called asparagus fern, to any other green plant you may have around, so if you cherish your cat, or alternatively if you cherish your plants and want to keep the cat off them, buy a small inexpensive pot of sprengeri fern and hang it low down where the cat can reach it; this should solve the problem.

With the end of this rather elaborate discourse one very familiar plant in sight, I hope it is clear that I do not dislike philodendrons—what I resent is seeing them ill-treated!

The Woods

During the past snowless weeks I took a trip up country, and on a cold dreary afternoon, for want of anything better to do, I went for a walk alone through some quite thick woods that consisted mainly of pines and hemlocks. I was no distance from a friend's house and not in the slightest danger of getting lost, and yet I felt a familiar curious sense of relief when the close-ranked trees thinned out and I saw open pasture land ahead.

I have always found thick woods a little intimidating, for they are so secret and enclosed. You may seem alone but you are not, for there are always eyes watching you. All the wildlife of the woods, the insects, birds, and animals, are well aware of your presence no matter how softly you may tread, and they follow your every move although you cannot see them.

I realize that there is something a little primitive and totally unrealistic about the slight sense of apprehension that always comes over me when I walk alone through deep woods, and I can't help wondering whether it stems from our ancestry. At some time in our most ancient history our forebears came down from the trees and walked out of the woods forever; and when that happened they became outcasts from a closed community; they ceased to be the watchers and instead turned into those who were watched.

Forests have a strong hold on the emotional history of every race; primitive people the world over believe that spirits inhabit forests. The old gods of the Western world constantly traveled through the woods, making their way to hidden sanctuaries concealed in thickets. In the subconscious of each of us, vanished religions remain half-buried in the forms of myth, folklore, and fairy stories, and in almost all the legends that have come down to us from our past, forests were hazardous, places of potential evil.

The silent forest is an equally sinister place in fairy stories. Red Riding Hood walked in danger because she had to go through the forest to reach her grandmother. Hansel and Gretel escaped from a witch's cage that was concealed in the farthest part of the woods, and at a very different date poor Mole was scared out of his wits by the curious noises in the Wild Wood.

As a child I always thought of the forests of these fairy stories in terms of the Black Forest. I imagined them to consist of dark whispering pines, to be soft underfoot, dimly lit even in the daytime, and not a place where it was ever safe to make much noise. Whatever atavistic traces may still exist in me from my ancestral past, there is nothing in them to suggest that the forest ever represented home and safety.

I sometimes wonder how the Pilgrims felt about those enormous forests of huge white pines that they found marching almost into the sea. In some ways the woods were good to them, for they provided food and kindling. But I don't doubt that those tall dark trees also produced unreasonable misgivings that had nothing to do with the obvious practical dangers of the shelter they afforded the local Indians who slipped so silently through them.

Snowdrops

Just before a recent brutal snowstorm I noticed that the snowdrops were already up and showing a gleam of white, but now, with more than a foot of snow in a single night, they have of course vanished completely and we will see no more of them for many weeks. Nevertheless, I like to think of them standing at frozen attention under all that snow, poised to cheer us when spring at last arrives, for the cold insulation will not harm those tenacious little flowers; it preserves them for our later pleasure.

As a child I had a secret with myself about snowdrops which may explain my slightly sentimental attitude toward them. There was a dell near where I grew up, a damp, thickly wooded, saucer-shaped depression in some open meadow land. And in it, crowding the open lower slopes, enormous snowdrops had spread into banqueting tablecloths of flowers. They appeared soon after the New Year, and in the cold damp springs of the English countryside lasted for weeks before they faded.

I don't think anyone ever visited that little hollow except me. To the owner of the meadow it was a nuisance in otherwise excellent pasture land, and it was fenced in with barbed wire to prevent inquisitive cows from tumbling down the steep sides. But to me it was a very special place, for apart from those amazing long-lived snowdrops, it was also always alive with birds and the small animals of the woods.

For some years I went there a great deal in my free time. I liked to sit quietly in it at any season and see how many birds and animals I could tempt by my complete immobility into going about their business. Because that neglected, damp hollow, with the shadowed light from the trees that clustered in it, gave the place an unusual atmosphere, it also became the enchanted center of the many fantasies I told myself. And since the bottom of the dell was invisible from the upper meadow, I constituted myself its unofficial guardian, keeping the slopes clear of weeds and encroaching brush and brambles.

I never told anyone about these visits, partly because most children love to possess a secret hiding place known only to themselves, but also because the land belonged to a neighboring farmer who still held to the rather rigid attitudes about trespassing that prevailed in the England of my youth. And I was well aware that had my parents known that I had made headquarters on someone else's land I would have been strictly forbidden to go there any more. For a time therefore that little grove of trees and flowers mattered a great deal to me, and I deliberately pretended it was magic ground where the usual humdrum rules of normal life did not count. In spite of my delight in those snowdrops I never picked a single one, for I had managed almost to convince myself that their unusual loveliness was a sign that they were indeed enchanted, and that once touched by a human hand this special beauty would vanish forever.

Indeed, unless my memory is entirely tinged with the nostalgic romanticism of childhood, those particular snowdrops were abnormally large and long-stemmed and quite unlike those that grew in the family garden. With my increasing age, any magical attribute must unfortunately be discarded, and some other reason must be found for their excellence. This may have come about because the growing conditions were exactly right for snowdrops, and the fact that the bulbs were undisturbed by picking would have led to the large increase. But I think there is also probably another explanation. The small snowdrop *Galanthus nivalis* which by the way may not in itself be native to England, is the species most usually associated with the name, but there does now exist in England in a half-wild state a larger snowdrop, *G. plicatus,* which is a Russian import that was reintroduced into England (for an earlier attempt had failed) by soldiers returning from the Crimean War. My guess would be that my magic snowdrops were originally Russian aristocrats that somehow escaped garden bondage and set up a happy exile in that cool damp dell.

In this country we cannot enjoy either the snowdrops or their close cousins the snowflakes (leucojums), which many of us also grow, for nearly as long a period as I did all those years ago. For with us much of their flowering time is during the weeks when they stand unappreciated under heavy snow. And when at last they do emerge, they often have a very short stay in full bloom in our unpredictable spring climate. Nevertheless I always feel sad when their flowers fade, for that marks the end of the first anticipatory delight in the coming spring, the tangible evidence that winter is finally over—something that is almost better in promise than in reality.

But though we can't all induce proud Russian aristocrats to take up residence with us, everyone should grow snowdrops, even the smaller humbler types, for their fragile appearance masks great sturdiness. As a family they and their cousins prefer a cool, damp place in light shade, although they are tolerant enough to survive almost anywhere except in full sun. Once established, snowdrops are not easily discouraged. They were some of the few bulbs that survived years of neglect in an empty lot next door to us, and in my own yard some have seeded themselves into the edge of the lawn. When the grass is cut the snowdrops are inevitably shorn of their foliage and should, according to the books, then perish. But each spring these outriders regularly reappear, spreading still farther into the lawn, flowering a trifle defiantly as though daring me to exterminate them.

Snowdrops can be bought in the fall as small bulbs, but planted this way they take a considerable time getting established. Both snowdrops and snowflakes are members of the amaryllis family and should, therefore, be very troublesome about being moved. But oddly enough better results come from digging into an established patch while the leaves are still green and the seed pods forming. Separate a big clump into several smaller groups, being as careful with the leaves and pods as possible, and replant the divided groups with a shovelful of compost or some store-bought soil underneath them. This is the only bulb I know that prefers to have established plantings broken up while the leaves are still green and which spreads faster with this handling than with conventional dormant planting.

So if you have snowdrops, mark them for division in May. If you have none, talk a friend into allowing you to dig up a clump: the chances are excellent that no one will ever notice where you have removed a few of these admirable little bulbs. After replanting, allow the leaves to wither and then forget them, and in the spring they will reappear full of flowers, with the delightful extra dividend of tiny green spears alongside which are the first leaves of the seedling plants.

And from then on, never fail to go and look for them very early in the spring when there is a temporary respite from the snow cover—for that is when you will appreciate them the most.

Clivias

Clivias, which are South African plants, are invaluable for the indoor gardener, though at first sight the price may seem a little steep. But since they dislike strong sunlight and zealous overwatering, and the arching strap-shaped foliage will remain a delight even in a hot, not very well lighted position, it would be hard to find a better present for the average apartment dweller.

Clivias are also delightfully immune to most of the afflictions that beset so many indoor plants. They do occasionally serve as hosts for voracious mealybugs, but this is a mild problem, for those pests vanish forever when touched with a cotton-wrapped toothpick that has been dipped in alcohol—either denatured or potable! Clivias increase in size unexpectedly fast. Well handled, they lose leaves very slowly; one or two may yellow off at the outer edge of the fan each year, but new growth emerges from the heart of the plant in greater profusion than this small loss, so clivias expand outward with dignified majesty a little like a mud slide that cannot be averted.

A healthy clivia also produces additional infant replicas of itself in the pot beside the mother plant. These too develop into big spreading plants even when they and the original specimen are still confined hugger mugger in the same pot. This multiplication of plants can thrive for an astonishingly long time under crowded conditions, but eventually the pot becomes all roots and no soil, and when that happens there are only two possibilities: either the offsets must be separated and potted up individually or the multiple plant must be shifted to a bigger pot.

I have seen highly attractive tubs crammed with enormous clivias, and if you have the space, the big lightweight pots that are now available make owning a gigantic indoors plant simpler than in the past. But there comes a time in most of our lives—unless we happen to be living in Versailles—when a point of no return is reached about moving a plant into anything still larger, and then division is the only answer.

This is not a difficult job, though it will be a messy process. A large sharp knife wielded with decision and precision is far the best tool. As long as the original plant and the new divisions emerge from the operation with plenty of roots attached to each individual section and are carefully repotted in well-crocked containers with fresh rich soil to comfort them, there will be almost no outward reaction to the ordeal except the possible yellowing off of a few leaves. The only question you will have to face is whether your available indoor area has space for three or four extra pots of clivias! And in the event that you can take no more, think of the reputation you will acquire as the donor of extremely generous presents.

Clivias kept in a cool bright place produce large umbels of long-lasting orange flowers held well above the foliage. The only problem with flowering clivias is the unpredictability of the blossoming period, which, when the plant is left to its own devices, can occur at any time of the year. Clivias belong to the amaryllis family, most of whose members loathe root disturbance. One of the reasons that proud owners prefer to move their plants into ever larger pots, rather than breaking them up, is because the trauma of separation, though it does not affect the foliage, does inhibit flowering, and a divided plant usually sets no bud for several years after the operation. Incidentally, even when the living is perfect a clivia will never produce more than one flowering stem at a time, although it may blossom more than once in a single season. Those pots you admire, filled with seven magnificent heads of bloom, indicate that there are seven mature plants crammed in together, and not that the grower has found a secret formula for getting more than a single bud from each individual plant! Once you notice a bud, a dilute feeding of liquid fertilizer is in order. But as soon as the bud shows color, cut down on the amount of water you give the plant and keep your hands off the fertilizer. Like all bulbs (and a clivia is a bulb even though it never loses all its leaves), the flowers will stay in good condition longer if the pot is kept short of water.

And if, when the plant flowers, you decide to become a surrogate parent, pollination can be easily done with a camel's-hair paintbrush. The round red berries that may follow last for months. I do not find that setting fruit lessens the likelihood of subsequent flowers. I have had many a clivia carrying berries all winter long and then flowering the following season.

If you don't want berries, cut off the flower heads after the blossoms fade, but leave the flower stem to wither away naturally. With all bulbs the slow demise of the flower stem plays an important part in building up sufficient energy in the plant to set the embryonic bud for the next year.

I happen to like my clivias to flower at the same time. They are more impressive massed, and I like that time to be the period when the forced daffodils are in flower, for the yellow of the daffodils combines well with the orange clivias. Clivias can be made to conform to almost any blossoming schedule that you want of them if you imitate the sequence of events that brings them into flower in the wild. Clivias in nature live on riverbanks, and are able to withstand long periods of drought when the stream levels drop during the dry season. The flowering period comes after the rainy season has raised the level of the streams, and given the plants renewed access to plenty of water. Home gardeners can imitate these conditions by withholding water for three months, during which time the pot soil can become as dry as dust. My plants go into this resting period in October, and the bud set is more reliable if the rest is given in as cool and dim a place as possible. The aim is to force dormancy on the plant, and this can be difficult in a warm room where there is a lot of light, which in itself stimulates plant growth. A clivia entirely deprived of water but left in an environment where it may still attempt to grow can be permanently injured.

Once the statutory period of rest is over, bring the plants back into a warm, light place and start heavy rewatering. But always make sure that there is excellent free drainage from the base of the pot. If you think about those riverbanks and the surplus water running off them, you will have no trouble handling clivias. Approximately six weeks after the rewatering has begun, a flower bud should appear on one side of the leaf arch.

Some years back, I noticed in Kew Gardens near London a pale yellow clivia which I longed to possess, and I started long, protracted negotiations to get an offset. For several years nothing at all happened, not even any answer to my letters, and then, through the intervention of a friend, I suddenly received a notification from Kew that one was available. To import plants into the U.S. from abroad it is necessary to get a formidable number of licenses from the U.S. Quarantine Department, and in spite of the long delay from Kew, I felt very sure that the plant would not remain available indefinitely if there was a slow reaction from me. I therefore sent a pleading letter to the U.S. authorities stressing the urgency of my case, and to my delight I received the necessary papers almost by return mail. These I sent at once to England, asking only, for it was then October, that the plant be not sent in the dead of winter. Once again there was total silence until out of the blue I got a notice from the local air freight depot that my treasure was there.

Naturally the notice appeared late on a Friday, just before a weekend that had a Monday holiday tacked on, and equally naturally the weather was bitterly cold, for the holiday was Washington's birthday. When I rescued the package I was pretty despondent. The storage area was unheated and the carton heavily damaged. But when, anxiously, I unpacked the plant I found it resting in a thick layer of dry excelsior, apparently unharmed, with the roots washed free of the least particle of earth. Potted up, it grew on all summer alongside my other clivias and looked so healthy in October that I put it away for the usual midwinter rest—something that is not always advisable with newly repotted plants. My optimism appeared entirely justified when it set bud the following spring, something I had not expected for several years, and produced a pale primrose head of flowers with a delightful faint scent.

But in one sense that was its swan song. That single bud was very nearly its last, and several years have gone by without any more, for the success story suddenly took a very serious turn. The summer after that entrancing flower, the whole leaf structure suddenly keeled over, and I discovered to my horrified amazement that, unnoticed by me, all those lower roots that I had extracted from their nest of excelsior had rotted away. There was nothing left to provide either nourishment or support for the arching foliage fan except a circle of small new roots that were just appearing under the crown of the leaves but were not yet long enough to reach into the pot soil.

As emergency action I took the plant out of its pot, cleaned off all the dead mess of old roots, and repotted the foliage mass so that those tiny little roots were below soil level. I used a much smaller pot and a great deal of internal drainage material and a very light sandy soil mixture in order to be as certain as possible that there would be no sodden earth in the pot to rot the bottom of the leaves. The invalid was then firmly supported between two stakes so that there would be no rocking of the foliage to disturb the little roots as they tried to establish themselves, and the pot went into a shady greenhouse where it could be carefully watched and watered only by a watering can—no more casual sploshing around with the hose. All this effort paid off, and strong new roots soon took over and re-established the plant. I have since been able to move the plant back into a pot in better proportion to the foliage span—but it has yet to flower again. The total collapse of the root system proved that there had in fact been serious damage to the plant while it waited in that cold warehouse, for although clivia roots do die out and renew themselves, normally they never all die at once. The flower that pleased me so much had obviously existed in embryonic form in the plant before it was dispatched overseas, for a preset bud in any bulb will flower if the growing conditions are even minimally tolerable.

I can't wait for that plant to prove that it is entirely recovered by flowering again. But I count myself lucky to have saved it at all, so I am not complaining.

Toads

One of the pleasures of resurrecting shared childhood memories with my brother is the rediscovery (or possibly the admission for the first time) of long-hidden feelings about some aspects of that life we once shared. My brother, being older and taller than I, saw things both figuratively and literally from a different perspective. Yet it is fascinating to discover how indelibly certain activities in my grandfather's house have imprinted themselves on our minds, even though we may recall them for different reasons.

Both of us, for example, loathed being sent into the conservatory in the dusk of the late afternoon. This often happened, for we would be ordered in to find some forgotten article. The conservatory was very large and contained wicker chairs and a round wicker table on which afternoon tea was occasionally served; and all our relatives had an infinite capacity for leaving things behind. My brother had—and indeed still has—an unreasoning fear of spiders, and before the days of modern insecticides, spiders were encouraged as useful allies against flying pests in any area that housed plants under glass. Late in the day, in dim light, for this was long before electricity reached the village, those spiders strung their webs everywhere in the conservatory, and he was terrified of brushing against one and feeling spider legs run across his face.

I didn't much mind spiders, and since I was too small even to see over the plant benches where most of the webs were hung, I was not likely to come into contact with them. My problem was a different one: what alarmed me were the toads that lived among the luxuriant ferns under the staging. The idea of unwittingly treading on a toad gave me the nightmares that haunted my brother about spiders.

Toads were encouraged in greenhouses for the same reason as spiders: they are natural predators that keep down the larvae of all kinds of destructive insects. And in an England of unscreened windows, the insatiable appetite of a toad for flies and mosquitoes gave it a particularly favored status. Any toads found at large in the garden were always carefully carried inside the greenhouse and encouraged to remain in residence by being served a saucer of water, sunk into the earth, which was kept constantly filled for their use. No one in those days knew anything about the territorial imperatives of animals and the need of every living thing to have a certain area reserved for its exclusive use, so what happened when too many toads for a limited area were let loose in the greenhouses I have no idea—perhaps they crawled away through the open vents. But I do recall my aunt rejecting an enormous toad offered her by the garden boy and telling him to take it back where he had found it. In answer to my inquiry she told me that old toads would never settle into new surroundings; to settle in, toads had to be caught young. Quite where she acquired this piece of useful knowledge I have no idea, but she established toads in every greenhouse she ever owned. These reptiles with their lumpy warts, yellow eyes, and pulsing throats both repelled and fascinated me by day, when I could see where I was treading. As privileged residents the toads never moved out of our way—it was up to us to make a detour around them—I used to stand and watch their flickering tongues as they caught a tasty snack. I thought them hideous and evil, a reaction no doubt to too many fairy stories. I used to wonder how any girl, no matter how kind her heart, could ever bring herself to kiss a toad, even allowing for the delightful possibility that an enchanted prince might suddenly appear in its place—something I dearly longed to have happen to me!

But time, alas, alters values. Today I have no wish to meet a transformed prince in the greenhouse; socially, I feel, the situation would be hard to handle. But I would love some resident toads to prey upon the pests, and I am not having much luck finding any, for toads have suffered enormously from our casual national overuse of hard pesticides.

Birds

Ever since the conservationists and our own observations have forced upon us the realization of the damage done to wildlife through indiscriminate spraying, we have all become much more conscious of birds and their worth in our yard, particularly those that stay around during the winter. In this area heightened interest in birds has led to a great proliferation of feeders: among my immediate neighbors all except one have a bird feeder, and a highly ornithological yard nearby now has three.

The effect of all this extra supply of food on the bird population has been delightful. There are now varieties of birds wintering over in this suburb that were never here formerly, cardinals in particular. The flocks of small birds seem to be suffering like the rest of us from a population explosion, all of which is extremely beneficial for the yard. For although birds are happy to freeload at the many feeding stations, particularly in bad weather, they also continue to go about their time-honored business of searching out natural food supplies, such as the egg masses deposited by various predatory insects, and in routing out hibernating mischief-makers from underneath the bark and in the cracks of the trees. Today I am sure no one needs to be told that the more birds a yard can support, the fewer insects there will be to trouble the gardener the following year.

But I am not the only member of this household interested in the increase in the number of the birds; it is also a matter of intense concern to our cat. We have always owned cats; I grew up with them, and the house would not seem properly equipped without one. But owning a cat and keeping birds supplied with winter food tends to lead to what medical leaflets term contra-indications, and in spite of our best efforts, there were in the past occasional tragedies. For when we were one of the few houses regularly setting out food, the birds came to depend very much on us, and there was often quite a problem trying to provide for the birds that fed on the ground and still keep them safe from the cat. For cover from the cold, and as a place for a quick strategic retreat, the feeder was put near a big hedge and some seed was strewn on the ground underneath the evergreens where it would not be buried by snow. But those same evergreens also provided a wonderful hiding place for the cat, and unless we were around to bring him in when the bird sentinels screeched warning, the ground feeders did not dare drop to the ground to take the much-needed food. I am afraid they were often terribly tantalized by seeing food and not being able to get it. For unlike a dog that will scare up a flock of birds and then rush away, a cat, even in bitter weather, will wait patiently for hours hoping to make a kill. And although you may tell yourself that the death of a bird is nature's method of eliminating the weak or the slow, I used to be upset whenever the cat brought a long cold stalk to what to him was a triumphant end.

But today all that is changed. With so many alternative feeding places the birds never need go hungry because the cat is hiding beside one particular feeder. When they spot him—which, since the present cat occupant of this house is yellow, they do very quickly—they just move off to another station. When he tries to stalk them there, they fly back to our supply.

Watching this happen day after day during the last cold spell, I almost felt some sympathy for the cat in his frustration. Indeed, when he gave up and demanded to come indoors, his air of bewildered disgust was so comic that I had to comfort him. But allowing the cat to sit on my lap when I really ought to work is a small price to pay for no longer having to worry about the possibility of his killing birds—or the nuisance of getting into boots and plunging outdoors in bitter weather to catch him so that the hungry birds can feed.

The Vine

During the past week in torrential rain mixed with, of all delights, fog, we made an excursion to the country garden to carry out some of those essential horticultural duties that can spoil the whole season's growth of a plant if they are done at the wrong time.

The main job was pruning the grape arbor, a relic from our predecessor in that garden that carries seven enormous old vines, all of which bear heavily every year. Unfortunately, to get good grapes in this area it is now necessary to spray the vines several times during the growing season. This particular crop is Concord grapes, and in a good year they are delightful. But since I am unwilling to indulge in such an extensive spray program for a nonessential crop, the fine preliminary set of grapes often withers or mummifies before the fruit can ripen. But even with this disadvantage the grape arbor remains a very important part of the yard, for it forms the shady sitting-out area for our guest house, and part of it also serves as the outdoor, slightly shaded area for summering many of the house plants.

But although I am not willing to spray heavily, I still try to give the vines every opportunity to bear properly, and one of the factors in getting a good yield of grapes is the manner in which the vine is pruned. Grapes turn into an unmanageable tangle unless they are regularly cut back and controlled, for unpruned grapes spend a lot of energy that should go into setting buds and ripening fruit, trying to climb up into the light and out of the tangle of stems of their own making. One of the most essential lessons that a novice can learn about pruning grapes is to be ruthless enough to let in plenty of light. This is best done by marking off five or six strong canes with plenty of new wood—that is, fresh growth from the past season, which will form the skeleton of the plant for the coming season. I usually mark my choices with a piece of red thread in several places along their entire length, for it is all too easy to cut one of the stems you meant to save. If the vine is trained against a wall, save only one central cane and two laterals on each side: if, like ours, it scrambles over an arbor, choose a number of young canes, which can be distinguished from much older growth by the fact that the bark does not flake away from them in strips. Once you have marked out those you mean to save, cut out all the others leaving only a few carefully chosen four-inch stubs to provide growth for next year. This produces a bare-looking vine, but it is an essential piece of husbandry if quality of the fruit is important to you. But still you have not quite finished the pruning, for once everything is out except the saved canes, these in turn should be cut back, leaving only about eight inches of the lighter colored wood of the past season as the source for the new fruit-bearing growth. Tie up what is left of the canes with loose twists of soft string: never bind grape canes in tightly or the flow of the rising sap may be checked.

For the home gardener that's all there is to grape pruning except that it is important to know the proper time of year for carrying out this drastic process. Here in New England the pruning date is some time in February, for grapes should always be pruned before the sap starts to rise. If you delay too long, sap will seep through the cuts, which looks and is messy, and can also damage the vines. In milder climates, grapes should be pruned around the end of the year.

So last week with frozen hands, numb ears, and water rising over my boot tops I set out to prune the arbor. It was a loathsome job and I got drenched in the process. And there had better be good grapes this year, for I picked up a dreadful cold during this activity.

March

John

He had been a sailor, and neatness was his passion and to a certain degree our despair. The sheds where he kept the tools were immaculate, with everything stored away by shape and size and tucked into obscure corners with baffling ingenuity so that we could never lay our hands on anything unless he was around to find it for us. This was not a situation that occurred by pure chance; he entered our lives rather late after he had been through a rough period, and he wanted the security of feeling that he was invaluable.

I would turn the corner of the drive on wet windy days and see him sweeping out his sheds, furious with the wind that had dared blow a few stray leaves inside. He was also an incurable pack rat. Nothing was thrown away and nothing went to waste. Coils of copper wire stripped of their insulation, chipped saucers, old tables from the dump, broken carving knives, and even a huge copper vat once used for rendering whale blubber were all carefully hidden away against that triumphant moment when someone might ask for them. He and I had a private joke that if I wanted a baby elephant he would be able to produce it—and he most certainly would have tried, for nothing was ever too much trouble or too difficult where keeping guard over this garden was concerned.

We worked together for nearly twenty years, and he always did produce—sometimes just an unexpected tool and at other moments that bubbling, lively interest that made everything such fun. Asked how things were going, his bright blue eyes would snap with pleasure: "Fine as silk," he would say.

He was a terrible worrier. When we were not around he took the responsibilities very hard, sometimes turning out late at night in the bitter cold for fear that something might have gone wrong. We begged him not to worry so much, but one of those extra visits on a hunch saved us from being devastated by fire.

He was, I think, at his happiest during the summer months when all responsibility was lifted from his shoulders by our presence and we could work together outdoors. He was supremely content during the years when he finally became assured that he meant a great deal to us, and when I was less busy and he was still strong, and together we remade a garden out of a wilderness.

To the last he was vivid, alert, and full of fight. He could always joke even when unhappy or in pain. He never lost interest in what went on, and deep down I half-suspect he hoped that somehow someday his damaged heart would heal and he could come to work again.

Recently in one sense we said good-bye to him for the last time. But tomorrow, or next week, or whenever we are in the country garden I shall remember him. He put so much of himself into that place that something of him will always be there. This I think would please him.

Salt?

I often hear about new gardening ideas just after I have treated some particular plant in entirely the opposite manner! Recently, for example, I repotted my regal geraniums only to read grave warnings in a gardening column a week later about carrying out this particular procedure at this season. I have grown good regals for years and am not about to change my ways, but it is interesting to observe that other experienced growers handle plants quite differently. Horticulture is such an inexact subject that I sometimes find myself offering the advice that if a plant is failing try caring for it in exactly the reverse manner. If you have it in bright light with regular watering, put it in subdued light and withhold water, and vice versa, for simple changes of this sort sometimes produce dramatic improvement. House plants all need light, air, warmth, and water in varying degrees which often only firsthand experience will teach you. But after providing these essentials, your treatment can vary very considerably and still produce good plants. This is as true today as ever it was in the past.

A kind reader has proved this by sending me a copy of *Ballou's Dollar Monthly* magazine for March 1857 which was published in Boston, Massachusetts, and despite its name sold for ten cents. In it there is a column called "The Florist," clearly compiled by an experienced gardener, and much of the advice handed out well over a hundred years ago is entirely relevant to indoor gardening today. Quite correctly, as we are well aware, we are informed that "parlor plants cannot be enjoyed in perfection without considerable labor and since plants breathe through their leaves and these must be kept clean, a light cloth should be thrown over them when the room is swept." The writer also notes that "some plants in flower will open their petals in the light of a lamp and close them again when it is extinguished"—surely a very early recognition of the importance of artificial light. There is also the familiar refrain that "plants accustomed to a greenhouse atmosphere do not thrive well when transferred to that of the common sitting parlor," as true today as ever, even though we now call that spot the family room. I was glad also to read that plants' "moral influence is refining even upon the youngest member of the family circle," something I wish my grandchildren would bear more in mind as they dash past the plant stand. So far so good; everything fits today's pattern equally well. Then unexpectedly this knowledgeable writer offers rather unexpected advice. It is strongly advocated that house plants should be watered only in the evening—which is not modern thinking—and even more surprisingly, "there is probably no plant that will not be benefited by an occasional moistening of the roots with a small quantity of salt and water, especially parlor plants."

Anyone writing so authoritatively must have had success with this remarkable treatment, and I intend eventually to try it on my own house plants—possibly, as a doubting Thomas, upon those I don't care too much about! But I do have some evidence about the effects of salt on outdoor plants that seems to suggest that there is indeed something to this advice.

Salt, as we know to our cost, overused on snow-covered roads is having a disastrous effect on many roadside trees, especially the sugar maples, which seem particularly vulnerable to it. The bad aftereffects of too much salt directly on the soil come as no surprise to horticulturists; many of us have used it in association with boiling water for weedkilling in preference to poisoning the land with chemicals. And in strong solution applied directly to the soil, salt is an effective killer. But it can also be used more selectively on specific plants by sprinkling their foliage with it and then misting the granules so that a strong salt solution spreads over the plant leaves and is absorbed into the cellular system. We have been using salt in just this careful manner to try to clear the root area around some azaleas of a new, dreadful infestation of ground elder. These particular calendula azaleas had been moved fairly recently, and I was afraid that their root systems might not yet be sufficiently well established to withstand either poison or strong salt concentrations on the ground. It seemed better to try to eradicate the ground elder and its insidious stolons by treating each sprout individually with a dose of salt as soon as it appeared above ground.

With a great deal of this painstaking individual effort we think that the ground elder is now beaten. But during the process I kept a very sharp eye on the azaleas, drenching their root spread with extra water after every "treatment" to dilute the potency of any salt that might inadvertently have fallen on the ground. In a way this was rather like taking coals to Newcastle, so wet was the summer, but I did not want to take a chance. And, interestingly enough, the azaleas that had the ground elder at their feet, and which must therefore have absorbed some extra salt in spite of our precautions, have set better buds than their relatives growing under identical conditions a little farther off.

Records

If you are an interested horticulturist of any sort—indoor, outdoor, or under lights—do start a garden notebook in the spring. This need not be in the least fancy; something from the five-and-ten will do very well, and your entries can be as brief as you like. But I wager you will be astonished at how fast those notebooks fill up, and the pleasure they will give you browsing through them later on.

I keep my old notebooks, which deal with every aspect of

growing things, in a bookcase in the cellar, and every now and then I go down and rummage in them to check on a faulty memory. And invariably if I start serious reading among those untidy books, the rest of the day is lost. Once begun, I cannot stop reveling in past triumphs—some of which now fill me with unadulterated astonishment and envy—and commiserating with myself over the failure of high resolves! If I could but find room for those books upstairs, and consult them more regularly, I would make fewer errors today. For gardening records of this kind, kept for more than thirty years, serve as an unintentional valuable teaching record in themselves.

I have kept garden notebooks since I had a yard of my own. I started the first, complete with black and white photographs, to show my parents the house and garden we had just acquired which they could not visit because of World War II. These old notebooks and the numerous boxes of colored slides of the gardens at their various stages which I also possess show our development as a gardening family and also throw an interesting light on many other half-forgotten matters: memories of past pets, for example, their names and where in the garden we buried them with formal honors. They remind me of the change in the bird population in the neighborhood, and they also provide valuable information about the unending vagaries of the weather in New England. Every year it seems to me I hear complaints about spring. It is either "late" or "unusually cold," "abnormally dry" or "fantastically wet," for no one is ever willing to admit that there is no such thing as a normal spring. Looking through my records it is quite obvious that although the weather may run in cycles, the variation of those cycles can be enormous. In an entry made in 1944 I mention April 21 as being a tremendously hot day on which I sat outdoors under apple trees in full bloom, and a passing reference notes that no doubt it was just this same sort of exceptionally violent early heat that troubled the British troops marching to Concord and Lexington. Three years later, on almost the identical date, there is an indignant mention of not being able to cut flowering branches to force indoors because the snow was too deep to wade through! I also have the complete planting record for the outdoor wedding reception we held in April 1963; we would have frozen into blocks of ice had we tried to do the same thing in April 1972!

We have always moved to the country in early June, usually to a rather backward garden and very overgrown grass—my books show that the mowers invariably collapsed about the end of May just trying to keep up with the grass. But in 1950 there is a delighted note: "front beds filled with huge stands of *Canterbury bells* in full flower, also Dropmore Blue *Anchusa* and Newport Pink *Sweet Williams.*" Somewhere along the line I seem to have lost the knack of growing these biennials, for most of my later records are very similar to the one for 1971: "Only one *Canterbury bell* survived the winter." Personal proof of the changeableness of our local climate makes it easier to bear the ups and downs of gardening!

The loss of the huge old elm that probably dated from the mid-eighteenth century and grew along our boundary line is now almost forgotten, for the entire planting plan was redesigned to hide the gap caused by its demise. But my notebooks show the enormous, costly efforts we made to save that tree and the near-despair I felt when nothing worked. I also recorded the insult-to-injury of what I felt was the exorbitant charge of four hundred dollars for taking it down, which included the cost for three days of a policeman to control the passing traffic. The charge today would run well into four figures and is another reminder, if we need one, of the way prices have changed.

I had forgotten until an old photograph surfaced recently that there once was a rose arbor in the town garden. But a check back to the notebooks told me where we bought the fir poles that still had bark on them and how we set them into laboriously dug post holes, and held them steady with stones that I dragged in from a deserted lot next door where the house had been pulled down. We planted a Paul's scarlet climber beside one post, and I diverted an old rose that I found planted against the house to climb over the top so we had a flowery canopy over the steps leading down to the cellar. For the life of me I couldn't remember what had happened to that arrangement until I read it had all come down when we built a small greenhouse against the back wall of the house and that in place of the roses I had planted a grapevine that I trained to the four-cane system with wire supports. It too departed when we enlarged the greenhouse. With my memory refreshed, I now recall my amusement, when I was harvesting a very fine crop of sweet green grapes, at being asked by some neighborhood children who had climbed over the fence whether the grapes were poisonous. I fear I intimated that they were very chancy eating unless you knew exactly what you were about when you picked them, for I was anxious not to have those luscious bunches pillaged.

Garden notebooks are instant nostalgia, and sometimes they can make you feel a little sad for times long since past. But if you keep them going and never let them become a series of faded relics they will form a continuing microcosm of family history as well as an invaluable horticultural record. So do start one of your own, don't allow it to become a nuisance, and don't feel that it has to be fine literature; write in it when the spirit moves you. This way you will preserve for yourself, and even perhaps for your children, a very pleasant account of how things were done by you and why.

New Ideas

I have just come back from making the rounds of a number of TV talk shows across the country on which—at impossibly early hours of the morning—horticultural questions are solicited by phone-ins. These are always rather toe-curling occasions, for considering how I myself feel live on TV at that hour of the morning, it is impossible not to harbor some grave doubts about whether the telephone will ring at all. Just before the "go" signal I am always torn between wondering why I ever consented to do any of this at all, or, alternatively, since I did say yes, why on earth I didn't line up a few resident friends to telephone in.

And when the phones do start to ring it is always a little humbling to discover how many people have a touching faith in the ability of a complete stranger to diagnose and cure the ailments of their plants—something that is in fact impossible without seeing the sufferers and knowing far more about the way they are handled. But the callers these days are not only concerned with sick plants; a far greater number want advice about growing things in general. Often the knowledge they show from their questions is impressive and bespeaks considerable gardening know-how, which makes this particular method of seeking advice all the more improbable. I have been asked complicated questions about soil structure, rotation of crops, organic gardening, and companion planting often enough to realize that many of my questioners are dirt gardeners with plenty of practical experience.

This was not always the case: I have not done much of this sort of TV work for four years, but I was much struck during the last venture with the change that seems to have come over the phone-in audiences. For one thing, the number of people plugging up the lines before the show even goes on the air is quite extraordinary. This has nothing to do with me in particular because it occurred equally in places where my TV gardening shows had been seen and where they had not. Whether the increase is solely due to a desire to hear one's own voice on air and catch perhaps an anguished look flash across the face of the guest as a tricky question is asked is not certain. To find that out it would be necessary to discover whether there has been an equal rise in audience reaction on other types of shows and with every kind of guest. But one thing is certain: large numbers of people who tune into these early morning talk shows everywhere are anxious to make their views known, and the viewing public interested in plants is far more knowledgeable and much more concerned with environmental questions than was the case formerly. Whereas once I was asked over and over how to make a Christmas cactus flower or what to do about African violets, I now have to field inquiries about herbicides, pesticides, detergents, and the possible overuse of nitrogen fertilizers as the culprit in the death of our lakes and streams. We may feel that we are progressing at a snail's pace in getting to grips with the matters that are so vital not only to our own health but to that of the next generation, but I am happy in the knowledge that the public is far better educated and much more aware of these problems than ever before.

From their voices, many of the questioners are also younger than those who called in a few years back; and it is the young voices who show themselves so much better informed about the conservation and environmental needs of this country. It is interesting also to notice that when specific growing questions are asked by youthful-sounding voices, these almost always deal with problems about raising vegetables or herbs.

We all know that one swallow never made a summer, and it would be foolish to make too much of this small personal sampling. But these seem to me to be excellent signs that the country is awakening at last to the narrow margin that lies between us and possible ecological disaster. Nothing will make me enjoy these sessions; they always have and always will intimidate me. But I have come back from this last round far more encouraged than in the past.

The Enclosure

A homeowner near us has just built a small enclosed entryway to protect his front door and has lighted it with glass walls that reach the ground. But to my regret, the structure was topped with a solid rather than a glass roof, and I feel a great opportunity was lost. For had overhead glass and some means of ventilation been installed, it would not have added immeasurably to the cost and would have turned that little enclosure into a miniature conservatory which could have been an inviting display area for all kinds of potted plants.

Even without heat, shelves along those glass walls could have been kept filled with potted plants for the greater part of the year; only in the coldest months of the winter would it have had to revert to being just a place to keep overshoes and boots. And with the addition of a little minimal warmth, either from an extension of the house heating system or from a small thermostatically controlled electric heating unit, it could have held winter displays of certain of the flowering azaleas, camellias, cyclamen, pittosporum, green bays, and flowing pots of ivy: all plants that can stand considerable cold as long as they have protection from hard, sustained frost.

This type of enclosure is not a greenhouse where plants are raised, but an exhibition area where purchased flowering plants can be kept in good condition for weeks. The exposure is not even particularly important as long as there is bright light. For although winter sun would be an advantage, such a small space would overheat in the summer if it faced south. If it faced north or east it could be a cool retreat filled with ferns, fuchsias, and flowering begonias. Sunlight is extremely important for bringing plants into bloom, but once they have reached that stage, they normally last longer out of the sun but in bright light. The little enclosure I watched going up faces east and is shaded by tall trees in the summer, and might therefore have needed the installation of some inexpensive banks of fluorescent lights for the maximum results.

My family had just such a glass-roofed enclosure built around the front door; one had in fact to go inside to ring the doorbell. This was not in the house where I grew up, but something that was added onto a smaller house my parents bought after my brother and I left home. When my mother realized that the one thing this new house lacked was a conservatory, she cajoled my father into having this tiny compromise built. And for the ten years they lived in that house she always kept her little greenhouse filled with a few choice plants that were a great pleasure to her, and which she could handle alone. This was very important, because the new garden was extremely large and the gardener could not be spared to help my mother with her potted plants. It would be equally important today when help to handle heavy pots is no longer available and most of us have no time to look after anything highly complicated.

My father used to complain that the plants were arranged in such a way that he invariably broke something whenever he walked through; my mother's rather unkind solution to this problem was not to rearrange the plants but to suggest that my father always use the side door instead! This not unnaturally added to rather than alleviated a basic problem, but in spite of the difficulty, they both took a great deal of pride in this rather ingenious idea of my mother's, and I think many of us could also profit from it here.

To have to pass through a little growing area before going into a house tells every visitor a lot about the family that lives there: plants are far more than mere decorations; they are indications of a life style. To be welcomed by thriving potted plants in a scented humid atmosphere sets a pleasant tone and shows a sensitivity and response to the natural world. The way plants are staged is also a means of expressing taste and individuality, something that we all need in these times of conformity.

Change

Recently I had the chance to take some plants from a large private greenhouse that was being dispersed. It was a sad occasion; the former owners had been enthusiastic, generous gardeners who had given pleasure to thousands over the years by displaying their flowers in plant windows where they could be seen by passers-by, exhibiting at flower shows, and willingly allowing visitors to walk through their gardens and greenhouses. The breakup of the estate was therefore an end of an era for all of us.

The available plants were wonderful specimens, but it was hard to find anything suitable for my much smaller greenhouses. Greenhouse gardening is now extremely popular, but, unluckily, it remains a luxury. The structure itself is expensive, the labor to erect it even more so, and the heat and essential self-operating equipment which an owner gardener must install if he is ever to be out of the house for an entire day adds greatly to the overall cost. In consequence, those of us fortunate enough to have a glass structure of any kind have to be content with something rather small, and this limits not only the number but the variety and appearance of the plants we grow. Very few new greenhouses, for example, can accommodate mature camellias or azaleas. Even though both these shrubs flower when small, there is a world of difference in the overall look of a mature plant and an immature example of the species. The same applies to many of the orchids, which are at their most spectacular when they are allowed to develop into stupendous plants in gigantic tubs.

Our modern peripatetic way of life often involves a second or vacation house, and this too has brought great changes to the material used under glass. If the greenhouse is to be closed down for several months it is not possible to grow climbing roses, plumbago, and jasmine trained on the walls and planted in the ground as I remember them in the family greenhouse. In consequence, here in the North we never see some of these plants the way they should look.

I love my greenhouse, and it repays me threefold for the time I spend on it, but walking through those soon-to-be-closed-down huge glass structures it was impossible not to mourn the coming loss of such gardens under glass, with the tall unusual plants none of us had room to accommodate and the great camellias planted in the ground that were too big and too old to be moved safely.

But there is no good mourning what cannot be helped. The important thing is to see how best to get the most out of the small-size greenhouses we can own. We have a greenhouse both in the town garden and in the country. Both are lean-to structures with the house wall forming one side, and they are both heated by forced natural gas that sends hot air through them with a fan that runs perpetually to give constant air circulation. A greenhouse in which the air is never stagnant can be run at a lower thermostatic setting than one without a fan. I do not, however, have nearly enough height to enable me to grow all the plants I would like. For as I and my specimen plants grow older, we all get a little larger, and the overall height even of the greenhouse I designed myself prohibits me from using tall plants on the benches.

For many years I kept things under control with steady root and stem pruning, but here I was getting to the point of no return. Even the kindliest plants can stand just so much of this annual mutilation, and I was getting very near the moment when I knew

I would have to give away some of my cherished big pots. But a couple of years ago I tentatively decided to try another plan. I abolished the back benches from the wall side of the country greenhouse and stood the biggest plants on the floor. This gave them ample headroom, and I used the space between them for plants that needed some shade. I moved with caution, for I was afraid that there might be a great drop in the humidity when the aggregate-filled bench was removed, but no ill effects appeared. This flexible arrangement proved a great success; the big plants did far better than before, and there was more space available for other pots. Now I have also removed all the benches except those on the sides in the town greenhouse and substituted cinder-blocks, piled one on the other, with planks laid across them for plants needing good light. And by putting narrow boards through the slots in the cinder blocks I have produced a three-tier staging arrangement that has more than doubled the space available for potted plants, and the arrangement has given me much more room for shade-loving specimens that have a hard time in a south-facing greenhouse when the weather warms up in late spring.

There has been a small loss in the overall humidity, but this has been easily compensated for by keeping shallow pans of water underneath the heating vents so that the hot air blows across them and carries moisture into the atmosphere. I have to keep those pans regularly refilled, so there is obviously heavy evaporation. Since all greenhouses are costly, and such items as benches count as extras not included in the packaged price, new owners might consider omitting them at first and seeing whether some homemade compromises aren't just as effective.

Tomatoes

March is a month of considerable frustration—it is so near spring and yet across a great deal of the country the weather is still so violent and changeable that outdoor activity in our yards seems light-years away.

By March, where I lived as a child, winter was really gone. We had the high winds of the nursery rhymes, but all the other vernal signs were there. Snowdrops were over; primroses and the tiny white anemones were thick underfoot in the woods; the daffodils were crowding out the withering crocuses, and the air was full of the cries of birds. In my present abode, the sounds of March are all too often the steady hiss of yet more snow against the windows and the scream of car tires spinning on ice patches. And although the birds call, their cries are the harsh ones of hunger. But in spite of all these apparent unending miseries, the days are in fact lengthening, and plants and seeds are aware inside themselves of a primitive urge to get going. They are programmed to respond to lengthening day hours, and they will answer this ancient call even under the most unpromising conditions. So March is an excellent time to get children interested in growing things: home horticulture begun now in a sunny window has a good chance of long-term success, for there is the real possibility that whatever is started now may be able to be set outdoors before ruin from indoor life sets in or the family becomes bored with the experiment!

On your next shopping trip stop off at a store and get a small bag of soil mix and a package of tomato seed; if you can find it, the variety Tiny Tim is a great family favorite. Then save and wash half-eggshells until you have refilled their cardboard carton with them, broken side up. The next job is to show children how to make two or three vital drainage holes in the bottom of the shell. This is not such a disaster-prone occasion as you might expect. Use a thin wire nail, an ice pick, or even a straightened-out paper clip (though its point is rather small). Press the sharp end of whatever you use against the bottom of the inside of the shell while it is resting safely in its carton, give the implement a slight twist as though using a screwdriver, and the point will penetrate through the shell without breaking it. Children love this job once they have been shown the method, and they are often neater-fingered about it than adults. But it is a good idea to have some extra half-shells stashed away so that no matter what happens you will end up with a full carton. When the holes are made, put the carton, eggshells still in it, on some sheets of newspaper and let your helper fill each egg with soil. A plastic teaspoon is excellent for this job. When each shell is filled almost but not quite to the brim, the soil should be watered. A mess-free way to do this is to use a well-rinsed squeeze bottle—an old detergent bottle is excellent, the kind that has a plastic nozzle with the end clipped off. After the watering, wait until all the surplus has drained out of every shell and then add a sprinkling of dry soil on top of what will be a wet mush—for the child never lived that was capable of using only a little water!

Now comes the moment for opening the seed package and having your assistant drop two or three seeds in each shell. If you are a really dedicated mother you can fill an old salt shaker with some of the bag soil and allow this to be shaken over the seeds until they just vanish from view. If all this is too much, have the child take up a pinch of soil between his fingers and thumb and sprinkle that over the seed. But make sure the seeds are not buried too deeply and that they rest on and are covered by unwatered soil straight from the bag; otherwise they may rot and never appear.

Keep the planted carton with the eggshells in it in a sunny warm window but don't give any more water for at least a week. Soon little seedlings will appear, and as they grow, the carton should be turned so that the plants won't bend toward the light. As an alternative you can have the child smooth foil over a piece of cardboard that is longer and taller than the carton and prop that up behind it. The reflected light from the foil will keep the seedlings equally straight. As the plants begin to grow, a few drops of water should be squirted on them each day, but don't let the earth get sodden. When the plants begin to crowd each other, give your helper a pair of nail scissors, choose which seedling in each shell looks the strongest and straightest, and snip the others off at soil level. If you are a great improver of every occasion, like all mammas in Victorian children's books, you can knock out one little eggshell load, show all the entangled roots, and thus make it very evident why thinning by pulling does harm.

In warm climates the eggshell seedlings can be set straight into the ground in a sunny place as soon as the weather settles down; but crush the shell carefully before planting. Roots will grow through the crushed shell, but they will have trouble breaking out of a whole one and the plant will be stunted.

If the weather is still cool after roots begin to show through the little drainage holes, plant each crushed shell individually into a pot of soil and keep them growing indoors until warm sun shines steadily. But whatever way these started plants are handled, whether they go ultimately into the ground or windowbox, or are moved into still larger pots, given bright and warm conditions they will eventually turn into fruiting plants which the children can harvest.

During loathsome March, this is a delightful way to get everyone going on a summer activity that will give long-term pleasure without calling for too much sustained interest.

Alice and Forsythia

This is the season when people lucky enough to possess a garden can start forcing out branches of flowering trees and shrubs for the indoors. As a child I remember long wands of various flowering bushes being brought indoors at this time of year. They were put in large dramatic groups in big vases in the icy front hall where sooner or later they got knocked over by my father. His outraged shouts about "your mother's confounded sticks" is a very early memory. I too get into something of the same trouble, for I like to put long branches where they sometimes snag the unwary—and some unremovable white water marks on the top of the piano bear testimony to those occasions when the vase got knocked over and I was not present to see that every drop of water was wiped up! But in spite of these problems, the dividend of early spring produced by forced branches is worth any slight inconvenience, and the time to start bringing them in is after there has been sufficiently cold weather to set the buds. If branches are brought indoors to force before there has been a period of prolonged cold, there will be no bloom, only leaves. I find it one of the few recompenses of a spell of bitterly cold weather to be able to go outdoors when the cold lifts and cut boughs with complete certainty that now they will accommodate me with flowers.

Forsythia is the simplest to force, and the yellow flowers will shine as brightly indoors as outside. But do make sure that there are plenty of squat, fat buds on any branch you cut; thin buds produce leaves, not flowers. Witch hazel, plum, quince, peach, and cherry all force easily; crab and orchard apples are more difficult, and lilac is a considerable test of the forcer's skill. The last three should be tried only near the natural season of blossom for the plant, and even with this precaution, I have had no success trying to bring branches of dogwood into flower indoors. Do not, however, think you have failed if pink flowers open white in the house; this almost always happens unless the forcing process can take place in an unusually bright place such as a greenhouse.

The simplest method for sure-fire success is to smash the final inch of the stems with a hammer and cut off any little twigs that might be under water level in the vase. The branches should be brought along slowly in the brightest but coolest place you possess, with the crushed ends in deep clean water. Lacking anywhere very cool, try forcing the branches in a bright bathroom where the constant humidity will help open the buds. Some experts recommend misting the branches; others suggest covering them with a thin plastic garment bag. I have never found any of this necessary except on rare occasions when I was forcing to a time schedule and wanted to hurry the process up. Once the buds begin to swell, the boughs can be brought to a warmer place and the small miracle of the unfolding enjoyed.

Forsythia forced this way has a very special meaning for me. My first winter in this country was long and bitterly cold, and I was desperate for spring, which I then was used to seeing appear far earlier. One day a new friend brought me an armful of forsythia branches still covered with half-melted snow—sensing my homesickness, she had denuded one of her bushes for me. I had nowhere cold and bright in the apartment in which we were living, so that forsythia had to be put in a hot, unlighted hall. But this particular present came to me late in the season and at a time when forsythia will flower even when forced under intolerable conditions.

And when at last in this strange country something came to life through my efforts I began to feel that here was truly home. Now each year as the forsythia flowers again for me indoors I remember that incident as the turning point in my feelings about this country, and recall with deep affection the sensitivity of that friend.

April

Potted Bulbs

Most of my potted bulbs were ruined this past winter during that spell of lasting, intense cold. Each year I plant in plastic pots—which I find crack less easily with the expansion of the frozen earth than the old clay varieties—a certain number of bulbs that I want to force later indoors. I set these pots in rows in three rather battered homemade cold frames, water them well, heap fallen leaves over them, then close the glass lids of the frames. We don't get back to the suburban garden, where this job is done, until late in the fall, and usually it is November before the bulb planting is finished. This is rather late in our climate, for bulbs need a considerable period to make strong roots in frost-free soil if they are to be forced successfully indoors later on. But with the use of cold frames and the accumulated mild warmth of the leaves, the pots remain frost-free long after the ground outside has become too cold for further rooting; by this method I normally grow excellent potted bulbs.

But if you are using the cold frame and leaf technique, there is one essential precaution. There must be no air space between the piled-up leaves and the glass top of the frame. If there is, the pots will not be sufficiently protected from the bitter weather of deep winter no matter how deep the frame, and the tops, which will have begun to sprout as the roots increase, will be frostbitten. The leaves we use for this coverup job are normally pretty sodden when they go in. Our late working date means that they have been thoroughly drenched by the fall rains. They have, therefore, a tendency to compact and sink as they dry out, and it is wise to check on them and add more if necessary about a month after the job is first finished.

Because I had to go abroad after this season's planting, I never made that necessary check, for snow had covered the frames before I got back. I found that the leaves had compacted considerably and that there was a wide air space between them and the glass in all the frames, and that is why I have no bulbs. If you have a really cool cellar with temperatures in the fifties, it is perfectly possible to do all this early forcing of bulbs there, just as long as the pots are kept in complete darkness. Lack of light is vital; with light the bulbs will sprout before the roots are strong enough to sustain them. Almost all the Dutch bulbs need about three months in near-total darkness, without the soil getting dried out, to make sufficient roots for successful forcing. I usually use both the frame and the cellar to force bulbs. I provide the darkness indoors, and save myself the worry of wondering whether the soil is drying out, by covering the pots with cartons with holes punched in them for air. Then I cover the cartons with lengths of plastic (opened-up plastic garment bags will do perfectly well) and tuck this in carefully under the bottoms of the cartons. This simple double play keeps the pots in darkness and preserves the moisture. Don't try to simplify matters by spreading dark plastic over the pots themselves. There has to be an airspace between the soil surface and the plastic; otherwise there will be mold.

I use both methods, because the cellar bulbs have a tendency to be ready for additional forcing simultaneously. No matter how hard I try, I cannot hold them back in the cellar after a certain stage of growth. In the cold frame it is possible to hold bulbs almost indefinitely once the weather is really cold, and bring in a few pots at a time—that is, if you can reach them; often we cannot because of snow. Having some pots in the cellar and others in the cold frame has always assured me of bulbs for forcing all through the worst of the winter. It was that same unexpected trip abroad that led me to forget about planting the cellar bulbs and got me into my present fix. And it is a fix, because I depend on bulbs for the display in the plant windows at this season and without them I am lost!

But the bitter cold did no harm to the bulbs that were planted in the ground, for the unending snow cover, maddening as it was, guarded them well, and it looks as though it will be a bumper year, which is some compensation for the indoor famine.

The *Iris reticulata,* a delightful, scented, miniature variety that has been produced for us by the hybridizers, is already out and doing better than usual. These bulbous iris have been hard to establish in spite of a lot of trying on my part. Normally they visit for a year or two and then peter out. This year, however, I notice that there is a big clump full of bud and bloom where I know I never planted them, so perhaps at last this picky plant has settled in. There are also clouds of the low ground-hugging specie crocus, an enormously prolific free-flowering group of small crocus mainly of the *chrysanthus* and *sieberi* species. But I am afraid they are waving their bright petals in vain, for the weather is still so bitterly cold that no bee in its right mind is going to leave the drowsy comfort of the hive to go nectar hunting in this wind!

Another mild triumph is the increase of flowers among the miniature daffodils. These tiny attractive bulbs flower before the large familiar types but are rather hard to grow. They are fussy about their living quarters and demand spring sunlight, bright light but not full sun, as the foliage withers, and a reliable supply of moisture in early fall, hard conditions for an absentee gardener to supply. Consequently they have not done well for us.

Then last summer the man who watches over the house and yard took it upon himself to trim off many of the ground-sweeping branches of the dogwoods that were my particular pride. This unauthorized mutilation infuriated me, for the hacking back was horribly obvious. But now it seems that I may have to eat the rather strong words I uttered when I discovered this unwanted pruning. For wouldn't you know that it is in this opened-up area under the dogwoods that clumps of miniature daffodils I had forgotten all about have reappeared with more flowers than I have seen in any of my other experimental plantings of this particular bulb.

Roads

This spring the main road I travel so constantly has been startlingly lovely, a slight recompense perhaps for the perils of driving on it during the just-past hideous winter weather! The trees that line the sides have outdone themselves in color contrasts, delicacy, and drama as the leaves unfold, and the pageant has been unusually long-lasting.

In New England we are very fortunate in the tenacity of our local woodland and native flora. When we slash through unspoiled country to make yet another road, our carelessness is forgiven far sooner than we deserve, for natural growth returns extremely fast and settles in compatibly with the outside material that is planted by those responsible for the upkeep of our highways. This is by no means the case everywhere in the country. In recent months I have been driven along a lot of superhighways from which all trace of the original vegetation seems to have vanished and been replaced only by rough grass. Whether this is a deliberate policy so that drivers can see as far ahead as possible without distraction or whether the original growth has been unable to make a comeback from the effects of the bulldozers I do not know, but these sparsely planted verges make a sad contrast to those we enjoy here. We are also fortunate that neither smog nor the depredations of plant hunters have ever suggested the feasibility of "planting" our roads with plastic reproductions which brought such howls of protest when they were introduced in California.

My regular route from one house to the other cuts across once prosperous farmland, and along it the wild and cultivated members of the apple family combine harmoniously. Father Bear, as it were, can still be spotted here and there in the shape of a gnarled old orchard tree, a solitary survivor, often in a factory area, but still loaded down with pink and white flowers. Mother Bear is represented by the various crab apples planted deliberately by the highway department in the wide spaces that occur occasionally as the result of the land taking. Baby Bear is the intrusive but ephemeral wild shadblow or serviceberry (*Amelanchier canadensis*) that, happily for us, refuses to be eliminated from its native countryside.

Other pleasing combinations are the groups of yellow-green willows planted after the road was finished in wet ground near swamp maples. Their colors contrast wonderfully with the russet red new little leaves of the maples. And both are ideally set off by stands of white pine, farther back on drier ground, whose dark boughs serve as a strong accent to the trembling fragility of the deciduous trees. This counterpoint effect is reinforced by massive plantings of gray-green Pfitzer junipers against the abutments of the bridges and the dark fingers of the native junipers that are slowly re-establishing themselves on the sunny, graded banks.

As it matures, this hodgepodge of local and cultivated material gives such pleasure that the original injury to the countryside done by the construction of the road can almost be forgiven. Woodland spring in New England has always been white, lime green, and red, set off by pointed firs, and the roads still maintain this basic tradition.

But although, with this careful landscaping, I can accept the appearance of some of our new roads, the problem of yet more tarmac monsters cutting swathes across what is left of unspoiled country is quite another matter. In our area a temporary moratorium is being enforced against what had become the accepted practice of displacing hundreds of people to drive new roads through heavily populated areas, but I have yet to feel assured that any similar moratorium on displacing wildlife and despoiling the environment is on its way. And the upheaval we create in the natural world by this insensitive road building has ramifications that, like the ripples in a pond, spread far beyond the area of the land that is immediately disturbed.

The papers are, for example, again full of schemes for still more roads to relieve the congestion of some of our major arteries. But roads beget cars, and cars beget more roads, and somewhere this circle must be broken. The damage done to the countryside by huge arterial roads cannot be measured only in terms of the loss of land from the housing developments and factories that inevitably follow them. Any road driven through open land also causes often irreversible damage to native wild flowers, birds, and animals. For not only are these disturbed and dispossessed during the construction process, but the aftermath in terms of their likely rehabilitation is questionable. The fill and fresh earth used to make the road and grade the verges usually come from a distance. Consequently they are often unsuitable for the regrowth of the local plants. I know a place that used to be full of the butterfly weed (*Asclepias tuberosa*), as vivid with its orange umbels as any

African clivia and now getting rare—it is in fact on the protected or endangered list. Clumps of this plant alongside a back road had somehow survived the dreadful period when herbicides were sprayed on the verges, and I used to make a special detour to enjoy them. But since the road was improved by being widened and straightened I have never seen this plant again. The truck-loads of earth brought in from afar that I watched being spread beside the new layout of that back road probably smothered the original stand, and in the new soil seedlings of that lovely plant could not re-establish themselves. And it is not only plants and shrubs that are often permanently dispossessed by road building; the area taken for the roads and cloverleafs also reduces the territory available to wildlife in general. For some years a newish road over which I travel had a sign DEER CROSSING where these animals, following an ancestral route to a nearby lake, had caused some serious accidents. The sign has gone now, and so has the danger, for no deer follow the old track today; not enough woodland remained beside the road to ensure the survival of the herd.

We must be grateful that our roads are so carefully landscaped and that native material is used wherever feasible; and presumably those narrow verge strips do house some of the displaced animals and birds. Recently, in fact, I read that since the new elevated roads had caused an increase in swamp areas, migrating waterfowl had sought these places out as resting grounds, and for such small mercies we can be thankful. But on the average we lose far more than we gain, for we know enough about the territorial needs of all living things to be aware that the land taken by the roads must lead to a diminution in numbers through attrition of hunting grounds.

Conservation is not an abstract term that does not concern us individually, nor does it relate only to immense projects like flooding valleys or cutting down the redwoods. Conservation is what is happening to the land in our immediate neighborhood, and its practice is how deeply we are prepared to involve ourselves, even at some possible personal cost, in protecting it.

The Dilemma

Our gardens are kept fertile and the plants in my greenhouse and windows thrive because I have access to unlimited supplies of rich compost which is made by piling up, turning over, and allowing to rot down all the leaves, weeds, grass clippings, and other garden trash that accumulate each season.

There is nothing particularly new about this; we have done it for years, except that in the opinion of my husband, who is the compost maker of this household, the time has now come to reorganize the area in the suburban garden where this work is done. Over the years my husband's skill at composting has become tremendously refined; he has brought the process to a fine art, and the finished product is beyond compare. But it *is* now years since he started this chore on my behalf, and although he is as energetic as ever, he is no longer willing to put up with the working arrangements in the suburban garden, which involved storing the finished product up a rather steep little incline. Instead he has decided to construct storage bins for this purpose on ground level close by the working area. And, alas, his chosen site, and indeed the only possible place with this new plan in mind, is the one area in the yard where yellow trout lilies have always flowered with unusual vigor.

Trout lilies (*Erythronium americanum*)—or dogtooth violets, an alternative common name—all originated in the New World with the exception of a single species. The small cyclamen-shaped flower with a mottled leaf is most commonly seen around here with a yellow blossom, although it does exist in other colors. It has absolutely no relationship to violets but is in fact a lily and a near cousin of the native American wood lily. I find the small yellow butterflylike flowers enchanting in early spring, but they did not find equal favor with the early settlers. One horticultural writer dealing with New England flowering rarities even went so far as to denigrate trout lilies as "bastard daffodills" (archaic spelling), which was extremely unfair—inasmuch as they do not bear the slightest resemblance to a daffodil except perhaps in color. The little flowering bulbs were far more appreciated in England, where they seem to have been introduced early in the seventeenth century and are praised at a slightly later date as being a pretty flower though lacking in scent. So maybe it is my English heritage that makes me delight so much in the appearance of their nodding heads and has led me over the years to guard the area where they thrive.

The books tell us that these bulbs increase easily from offsets and seeds, but this has not been my experience in our yard. In the past I have made several attempts to establish new colonies in suitable positions but always with marked lack of success. The only flowering subcolony that does exist sprang up spontaneously some years ago beside our back terrace. And although it has flowered, the blossoms are very sparse and nothing to compare with the thick stand near the working area from which it must somehow have originated. As with all lily bulbs, these never go completely dormant, and they are, therefore, extremely tricky to move. The best time for transplanting seems to be just after the leaves have withered, but since they are extremely small they are almost impossible to find at that stage. If you are attempting the job it is vital to stake every bulb very carefully while you still can see it. The only problem here is that a stake driven in exactly beside the bulb may injure it beyond recall, while a stake put into the ground a little distance away may not be close enough to enable you subsequently to locate the bulb. No horticultural manual I have read makes the point, but it seems very likely that trout lilies, like so many other wildlings, need compatible soil bacteria around their roots to thrive. So it is extremely important to lift them with as much earth still clinging to them as possible and move them into growing conditions exactly like those they inhabited before, with extra earth from their original position spread around them. This is just what we have been doing for the last two springs, ever since my husband decided upon the change, but our chance of long-term success has, I fear, been lessened by the fact that we have had to dig these bulbs while the leaves are still green, for we are not at that house after the foliage withers.

During this operation, in an attempt to be totally conscientious, we have not only lifted each bulb with a shovel so as to ensure the maximum amount of original earth, but we have also spread soil from the old site underneath it before it was replanted and on the ground above it. Since we have moved several hundred tiny bulbs, this has been a considerable undertaking; but I hope that with such a wholesale transfer of soil, if the bacteria that inhabit it are a factor in the long-term success of trout lilies, we have transferred enough so that they will in turn inoculate the ground around them with still more of their own kind.

It will be three years at least, judging from my earlier efforts, before we can be certain whether our attempt has been successful. Plenty of small leaves reappeared this spring from last year's work, but that is not a sure signal that the bulbs have settled in. Only the appearance of flowers will be convincing.

I have watched and assisted in Operation Trout Lily with ambivalent feelings. Obviously the compost maker must be allowed to lighten his labors, and yet, deep down, I have the uneasy feeling that despite all our care, my much-loved and long-tended stand of these delightful little flowers may show their resentment "in the answerable way of plants with the simple protest of death."

Easter Lilies

Whenever I see pots of lilies offered for sale all prinked up in Easter finery of ribbon and colored foil I get depressed. This I know is an unreasonable reaction to the pleasant custom of giving friends plants to mark a happy season. But so many of those lilies are doomed to such an unnecessarily short life, and they are offered to us in a denatured form which I resent.

This has nothing to do with the tawdry wrappings that can easily be removed. My objection is a basic treatment most flowering Easter lilies undergo before they are offered for sale: their pollen-bearing anthers are removed. Lilies look far lovelier to my eyes with the trembling sacs that carry the pollen inside the throat of the lily still intact. When they are removed, the flower becomes a waxen effigy, almost as emasculated as its plastic counterpart.

The official explanation for this action is that the pollen sheds freely and stains the petals. Purchasers are said to dislike untreated plants because they look "messy." But the cost of a lily is measured by the number of flowers and buds it carries, and since the unopened buds still have the pollen anthers intact this is not an entirely valid argument. It suggests that the retailers are willing to treat the flowers that the unsophisticated buyer can see, but leave them to discover for themselves that the buds which drove up the purchase cost will be "messy" when they open.

This, I am sure, is not the intention of the retailers; they have no wish to deceive their customers, and it would be better if the real reason for taking out the anthers were better understood. Raising lilies for the commercial market is a huge undertaking and a very chancy one. The plants must be timed to an exact date. Easter varies from year to year, and the flowers must approach peak at that date no matter how difficult the past few months may have been in regard to light, warmth, and so on. For just the right appearance, the lowest flowers should be open and the buds show a glint of white, the condition that most attracts customers. Inevitably in this difficult job some lilies open rather earlier than they should—anyone who has ever tried to force flowers for an exact date knows that it is better to have the flowers a little advanced, for plants can usually be held but not hurried. To keep the opened lilies in good condition, the growers lower the temperature in the holding greenhouses, and they also remove the pollen sacs. If these were not cut out the flower might be fertilized, and a fertilized flower, its mission accomplished, fades much faster than one that is untouched. It is therefore extremely important to the commercial growers of lilies that the prematurely opened flowers are infertile. That, not the so-called messiness, is the real reason for this form of plant castration.

This is a perfectly valid action where the growers are concerned; I just wish that the true reason for it were more openly acknowledged. For if purchasers knew this fact, they could make a better judgment about the length of time a flowering lily has been in bloom. If there are two or three buds fully opened with the golden anthers all removed, the lowest flower is already quite old. You will get a better bargain by buying a plant with one flower open and the other buds just unfolding.

I have great sympathy for commercial growers, and I cannot express often enough my admiration for their skill in timing and forcing the perfect plants they offer us. But from the purchaser's point of view, I also wish some of the effort that goes into removing those anthers could be used instead to provide better growing conditions for the lily bulb inside the pot. If the internal conditions inside the pot are reasonably good, potted lily bulbs can give you long-term pleasure with some rather simple aftercare, so it is important when you buy the plant to make certain that there is proper internal drainage. You can check this out by pushing a pencil up through one of the bottom drainage vents. If the pencil strikes something hard, there is a layer of stones at the base of the pot which is providing adequate drainage. Such a plant is a good buy. If the pencil goes up and up into soft earth, the bulb has not been grown properly and you would do well to look further.

After the lily that you have bought with all this forethought and care finishes flowering, cut off the blooms but leave the actual flowering stalk to wither away naturally. Keep the plant growing on in a sunny window with a reasonable amount of regular water. With lilies—indeed with all house plants—it is always wise to stand the pot on a pad of pebbles in the saucer underneath it. After watering, empty the saucer of any surplus water that rises above the level of these stones. Lily bulbs rot easily, and constantly wet soil triggers off this trouble. A lily kept growing on indoors which is slightly underwatered, that is with the surface soil allowed to get quite powdery to the touch, will do better than one in which the surface soil is always wet.

Some time after Memorial Day, when the weather has settled down, dig a hole several inches deeper than the pot itself in a well-drained place in the yard. Either full sun or deep shade is a poor choice, and a place where water stands will be fatal. Morning sun with good air circulation is best, but if you have to choose between no sun or too much, take the sunny position and plan to spread moisture-preserving mulch on the soil around the lily after planting. Put a pad of stones (similar to that you had in the saucer) in the bottom of the planting hole to a depth of about two inches. The old advice was to use sand as the base for the lily to rest on, the idea again being to provide quick drainage immediately around the bulb. Now, however, lily experts suspect that sand holds too much water, and stones are suggested to form a miniature dry well. Don't add fertilizer; lily bulbs are extremely sensitive to any of that against their fleshy scales. After thoroughly watering the pot, knock out the plant, keeping the pot soil as intact around the bulb as possible, and center it on the pad of stones. Lilies are some of the few plants in which it does not matter if some of the stalk is buried—indeed some lilies throw roots from buried stalks. Don't overdo it, but remember that all lilies except the lovely *Madonna* prefer to be planted outdoors more deeply than is possible in a pot. While you can still see what is going on, drive a stake into the side of the opened-up planting area without injuring the bulb, then fill in the hole.

The old foliage will soon wither away and the stake will remind you not to go digging in that particular spot, for sometimes new leaves and even a flower stalk will appear that same fall. But if this does not happen don't despair; leave the marker in and wait several weeks after the other perennials are above ground the following spring before you decide the lily has died, for these are plants that often are slow to break ground.

Maybe you will never see the plant again, which is a pity, but you will have lost nothing. More likely, however, the lily will resurface and flower and give you all the special pleasure of renewed acquaintanceship.

Ivies

To try to make some sense of the behavior indoors of various types of ivy, a plant that does not live comfortably with me, I recently accumulated a considerable number of pots of different varieties and ran some rather simple tests, which in effect amounted to trying them out in the house, in the plant windows, and in the greenhouse under various intensities of light and with different patterns of watering. I cannot pretend that I discovered anything the least earthshaking, though I did decide that daily misting with lukewarm water is of paramount importance wherever the plant is grown, and that ivies prefer rich moist soil and rather loose potting—and that, in my case, aphids indoors are apparently inevitable! I continued the experiment through the summer, taking the pots outside, which led only to the familiar conclusion that ivies continued to need plenty of water and preferred light shade.

By fall I had had enough of babysitting what are, to my mind, extremely unresponsive plants, and since I had proved so little of any real interest, I decided to give the whole thing up. But those plants had been through a good deal at my hands, and I didn't feel quite right about throwing them out. Instead I planted them all, those hardy and those alleged to be tender, in rich soil underneath a crab apple tree to take their chance with our tricky winter and see whether they would form the nucleus of a new groundcover bed. One of my pet peeves with ivy, as a ground cover used extensively, is its dullness. That dark green billowing uniformity in many California front yards is extraordinarily uninteresting. So this seemed a good chance not only to find out which types were in fact hardy, but also if an assortment with varying shapes and sizes of leaves would look more interesting.

The snow came early that year and in quantity, and the ivies, some of which were in poor shape after living with me for a year, vanished in early December under a snowfall of almost prehistoric proportions, not to see daylight again until April.

During the cold months I thought of them occasionally and blamed myself that through sheer sentimentality I clearly had let myself in for a mammoth job of spring replanting—it would have been far simpler and much more reliable to have set out tulips. But when at last the deep freeze ended, the ice age unlocked, and those ivies resurfaced, I found to my considerable surprise that all of them, even those that were such miseries when they were set out, had come through the winter in fine shape with shiny undamaged leaves and not an aphid in sight! One miniature variety, usually confined to ivy topiary work, appeared to have been so happy buried like a prehistoric mammoth that it had begun to scramble up the trunk of the crab apple to get away from the smothering leaves of the other varieties.

The reason for all this abundant health was of course the unending blanket of deep, insulating snow that allowed filtered light to reach the leaves while providing protection against bitter wind. For it is not the cold in itself that kills the ivy stems and burns the leaves, but cold wind that extracts moisture from the foliage which the frozen roots cannot replace, a simple horticultural fact of life that I had temporarily forgotten.

In this area we cannot be certain of deep, winter-long snow, something for which my nonhorticultural side is profoundly thankful. If we want to grow flourishing ivy outdoors that is not a mass of dead leaves each spring, we must provide a substitute for snow that will shelter the plants against wind and yet let in air and filtered light. The answer is a thin blanket of leaves raked over the ivies around Thanksgiving, but preferably not maple leaves, which pack down into a wet, suffocating mat. Don't spread these thickly like a mulch; just use them as a light fluffy covering through which you can spot the plants. If there has been no snow by the New Year, cut boughs off the Christmas tree can be added as extra protection and to prevent the leaves from blowing away.

A very important point in preserving the good health of an ivy ground cover in this climate is to remove the covering slowly. Snow takes its time vanishing, and if you act in the same manner it will help the plants. More ivies are killed in the strong cold winds of late spring than at any other period, so don't be lured into whisking off the protection during a premature hot spell in March. Leave the evergreen boughs in place until April. Unless the year is very abnormal, the branches will still be green and pleasant-looking. After the boughs are off, don't touch the leaf cover for at least a week, and wait even longer if the weather is still bitingly cold. I take off the leaves when the small specie crocuses begin to show color.

The leaves should be got out by hand. This admittedly is a back-breaking, time-consuming job, and it is very rare that you can kneel to do it on the wet ground of spring. But raking ivy beds invariably injures the growing tips of the stems; once these are damaged they take forever to make new shoots. During the cleanup, pick off any withered stems or dead leaves.

A great many people are so devoted to ivy that this will seem a small price to pay for its much better appearance. I don't love the plant with all that passion, but I do use it extensively outdoors. And I have now carried that particular test bed very successfully through four consecutive winters with varying amounts of snow. All the varieties are still alive, and the different leaf patterns do give the planting better texture and a much more lively appearance. I have to trim back the larger-leafed types—spring and fall—to prevent them from overwhelming the miniatures, but with that one caveat, the experiment has been a great success—though I *still* grow disgraceful ivies in the house.

The Pond

Every spring as I slosh across the flooded back lawn in the country garden, I glance with regret at a small area to one side now thickly planted with berried shrubs and bound together with wild roses. This was an abandoned water garden when we first saw it. It contained a sinister sheet of black ice covered with leaves and scum, and there was a carved stone head of a lion, once a fountain, peering menacingly—and perhaps a little forlornly—out of an engulfing tangle of bare branches of bittersweet, brambles, and poison ivy. Some of the concrete sides had fallen in, a small maple had established itself on a central island, and the whole area was overhung with tall trees. Wide puddles of black ice around the dilapidated verges showed also that the excavated area no longer could contain all the water that accumulated in the neighborhood.

I had grown up with ornamental ponds; they were rather a specialty of my mother's. She correctly maintained that plenty of active goldfish kept down mosquito larvae and that the frogs, which invariably settle in where there is permanent water, ate the egg-laying mosquitoes at an earlier stage in the perpetual circle of natural controls that exists where nature is left alone.

I had loved those ponds, which my mother made wherever she gardened. The birds used them as drinking fountains by day, and paw marks in the nearby mud showed where the small wood animals used them for the same purpose by night. Since I had already learned to respect the aggressive voracity of New England mosquitoes, I was strongly in favor of reactivating the fish and frog method of natural control as fast as possible! Our efforts were slowed by terrible afflictions of poison ivy, but eventually we grubbed out the invaders, patched and rebuilt the sides of the pond, and I set to work to plant a water garden around it. Once the water itself had been recontained, some of my predecessor's plants reappeared around the verges, giving me a fine head start. An extremely old-fashioned, small, double daffodil pushed up in quantity in early spring and was always the first to show color in the yard. Big patches of lily of the valley and what I inaccurately call wild lily of the valley, but which is technically known as *Maianthemum canadense,* or the Canadian mayflower, also appeared. And soon sheets of foam flower—*Tiarella cordifolia* for purists who like to have the names correct—moved in. As our contribution we undertook the enormous job of heading back the overgrown trees that had spread right across the pool. Any water area in a garden is much less trouble if it is not necessary constantly to fish out leaves—something swimming pool owners have also discovered. If you want to grow good water plants and get fine flowers, strong sunlight is also essential. The neglected little pool was so positioned that it was impossible to provide sunshine all day long, but by felling several ratty swamp maples we managed to open it up completely to the east and south, and a very thorough thinning of the remaining trees produced light-dappled shade in the afternoon.

We also pruned the maple tree on the island to a few angular branches. We couldn't grub this out entirely because that would have endangered the bottom of the pond, but since I was in a Japanese phase at that time, I hung innumerable stones from cords on the remaining branches to try to train them into interesting shapes. This rough and ready oversize bonsai method can be very successful with pliant young trees. The stones (which we encased in onion nets) look a little odd at the time, but the final effect can be quite exciting.

We added clumps of Siberian and Japanese iris and tall plumed grass along the sunny verges; on the lightly shaded side we planted astilbes, hostas, and aconites. Into the pond itself in April we sunk baskets, each planted with a single hardy water lily. This is the one planting occasion when compost is not a suitable growing medium. Some kind of coarse fibrous loam is better. To prevent all the soil from washing out of the basket when it is lowered underwater, spread a thick layer of sand on the top of it. At first we used to winter those baskets over in the cool cellar, lifting them out in late November, for I was not entirely certain that the plants would survive the bitter winters. But they didn't survive well in the cellar either, and in the end I left them all year long in the pond, where usually they pulled through. These days there are new dwarf varieties of water lilies available that can be sunk in clay pots. These would have been much better for our little patch of water than the larger-leafed varieties which we then had to use. And true to my intention, I released goldfish. But these unfortunate fish had to be preternaturally agile to survive, for a neighbor kept large flocks of half-tame wild ducks and Canadian geese who considered our fish a gourmet's delight whenever our backs were turned. Between the birds and the fact that the pond, even after we cleaned out most of the accumulated silt, was not really deep enough to allow the fish to survive in extremely cold winters, there used to have to be an annual ceremony of restocking from the five-and-ten. This took place every spring after I had frightened away the predatory geese sufficiently and often enough to make the water safe.

As usual, countless frogs appeared from nowhere, and the children used to go through the ancient ritual of scooping up the spawn and watching the subsequent development of tadpoles in a fish tank. But above all, we had bullfrogs of character and distinction, for no matter how much the ducks and geese ravaged, the bullfrogs always managed to escape. In early summer every evening and at dawn's early light they bellowed their lovesick cries. It was always a bit of a toss-up whether to allow yourself to be annoyed by the noise they made and think back longingly during those weeks of courtship to the relative silence of the suburban garden, or to be proud of the fact that no one else had bullfrogs that could match ours in carrying power or operatic fervor!

When the 1954 hurricane struck it was accompanied by a three-foot tidal wave that surged across the entire yard. All that weight of water had some permanent effect in the underground spring that fed the pond, for from that time onward it was never possible to keep water in it. Even more oddly, the spring floods or the summer flash floods following thunderstorms no longer drained into the pond but spread instead over the nearby grass and eventually found their way into our cellar.

For a time we kept the pond going; we restored the sides which the hurricane water had rebroken, and as a final act of despair we even put in a cement bottom and tried to keep the water level up with hose water. But this did not work, for when hot weather arrived the town invariably limited the use of outdoor water and the fish and plants died. Unfortunately, at that time I did not realize that there was an old well shaft hidden away under one of our large trees. Its presence was revealed to me only several years later during a visit by one of the daughters of the original owner of the house. We now never lack for water, and possibly we could have kept the pond going with this supply, but I am not even sure of that, for as an aftermath of that hurricane the pond drained of water continuously no matter how much was poured in.

Finally we gave up the struggle. We broke up the cement so laboriously laid down and planted the area deliberately as a bird sanctuary, setting a small summer feeding stand nearby and putting a hollowed rock on the ground which we keep filled with clean water. Water is as important to birds in the summer as food is in the winter and is an excellent way of keeping them in your yard.

There's no gainsaying the attractiveness of the area today; it is full of bright flowering bushes and their berries. It is a success as a safe retreat for birds, and, unfortunately, it is also a safe retreat for mosquitoes. For though we are scrupulous about allowing no stagnant water to collect where these pests can breed, they still stick to their old haunts and manage somehow to deposit their larvae with great success. And since there are now neither fish nor frogs to control them, I sometimes wish I could enmesh myself in mosquito netting before entering the area to load up the wheelbarrow with compost that is stored nearby. But I now have no doubt at all about one thing. When we come down here in the spring while the ground is still very wet I would far prefer to be able to hear again the cries of the bullfrogs rather than the endless moans of the sump pumps as they struggle with the ever rising water level in the cellar; which is where the ghost of the lost pond reappears each year.

Heaths and Heathers

In spite of much excellent advice to the contrary, it often happens that garden reorganizations come about through some rather minor changes rather than through pen and paper plans. A case in point is our new big heather bed, which I arrived at almost by chance and which I could not be sure had succeeded until this month.

A few years ago some structural alteration to our guest house forced us to take out a number of quite big foundation shrubs, and when the work was done, we found that the changes in the building made it impossible to reset those bushes in their original positions. This left us with a long narrow empty area which promptly became a long-term nuisance.

First I tried to turn the empty site into an herb garden, for the position against the house near a kitchen and facing due south made this seem a brilliant idea. But since the guest house was in constant occupation—and guests should be allowed a little privacy—I could not give the herbs and the bed the necessary attention, and lusty crab grass moved in and overwhelmed everything. Next I planted low-growing perennial asters that make a tremendous weed-suppressing billowy show elsewhere in the yard. Unfortunately, the year these were set out proved to be unusually dry, and the rooted cuttings did not get enough water to take hold fast. By midsummer the inevitable crab grass reappeared, accompanied this time by sneeze-inducing ragweed, and ground elder or goutweed, the worst pest that can be let loose in any garden. The appearance of the ground elder called for immediate, violent action, for every scrap of root of this plant that remains in the ground sends up new flourishing colonies. Unchecked, it can ruin a garden faster than any plant I know. To clean out this pest, we removed down to hardpan all the soil in that long strip of land which lay between the wall of the house and a concrete terrace. We put the infested soil into plastic barrels and carted it off to the dump. Our town operates a clean, deep, landfill dump operation, but I suspect that a green haze of ground elder will eventually work its way above ground no matter how deeply it has been buried, and when that happens, no one will be more innocently surprised than I.

With visitors again on their way to the guest house, the cleared trench looked so bare that I decided to brighten it up with some potted chrysanthemums. But when I went to our local nursery to get them, my eye was caught by a "sale" sign beside a large group of potted heathers, and in these plants I suddenly saw a solution to that problem area. Well-established heathers make a thick, spreading ground cover that smothers out most weeds and, as an added bonus, provides delightful long-lasting flowers from July onward. The plants need very little care during the growing season, and the flowers continue to look reasonably respectable after they fade—what could be more suitable for a planting around a guest house that had of necessity to be neglected?

In an extravagant gesture I bought a considerable number of these excellent plants in a color range that ran from deep purple through two tones of pink to pure white, and carried them home in triumph. But with that same gesture I let myself in for some hard, hasty labor. Heathers do not like standing around in paper pots with their root balls insufficiently protected from searing sun; they are, after all, plants from cool uplands, and blinding sunlight beside a hot road is very unlike the misty moors of their homeland. Those I bought looked as if they had already suffered too long from their fate.

To do their best in this country, heaths and heathers (for

amateur gardening purposes the two plants are interchangeable) need moisture but excellent drainage, a warm, sunny, sheltered position, and a sandy soil with an acid content. To provide for these requirements, the empty trench had to be edged with bricks to create a raised bed for the essential drainage. Filling it was extremely exhausting work. An enormous amount of soil had to be brought in, of which the component parts were rich compost, sand, and damp peat moss. And each barrowload of earth that was tipped in had to be trampled like grapes in a vat to prevent later shrinkage, because we could not wait for nature to do the compacting. The bricks enabled us to raise the soil about four inches above the level of the ground, and once the fill was in and stomped down, the heathers were planted. This took all day, for I wanted a good gradation of color and height, and the potted plants were moved around a lot before they were knocked out and got into the ground. The new fill was pleasantly moist and no extra watering was needed, for I firmed the soil extremely carefully around each individual root ball. But there was another job ahead, which was to preserve this moisture. This was achieved by laying down a heavy mulch of fir bark, which added to the acid content of the soil and did an excellent job of conserving moisture and holding down incipient weeds. It also provided an excellent foil for an alert watch against the much-to-be-dreaded possible reemergence of ground elder, which fortunately did not happen.

But in spite of all the effort the work was worthwhile: the bed took on immediate style as well as being slightly unusual. Heathers, being clannish plants, look their best massed together; anything extra planted among them weakens the effect. But although I was extremely pleased, I knew I was far from out of the woods where that new planting was concerned, for heaths and heathers are not entirely reliable in this area unless there is a heavy snow cover. I have grown them successfully in town for years, but there I can rely on plenty of insulating snow. The problem with this new planting was the occasional scarcity of snow in the country garden. And the question in my mind was whether the slight leakage of heat through the walls of the house against which they were banked would compensate for this lack.

As it turned out, we had a long mild fall which kept the ground free of frost almost until Christmas and enabled the plants to make strong new roots. But I still was a little nervous about the foliage. Ideally heathers should be set out in early spring, allowing the roots and the foliage time to toughen up. This obviously could not happen to the new planting, so the survival rate would depend not just on those new roots but also on the ability of the foliage and tiny wiry leaves to take bitter wind if there was no comforting snow to engulf them.

When hard winter weather eventually came, it was not too bad by some past standards, but there was extremely little snow in the country garden and an abnormal amount of wind—exactly the combination most likely to prove fatal to those plants. After the third howling gale in near-zero weather, we put up a windbreak of snow fencing to filter the wind and piled leaves over the plants, holding them down with chicken wire. I had not wanted to do this earlier for fear of burning the foliage, for when the wind is not blowing, their growing position is a sun trap; between these various choices of evil, I was not optimistic about their survival.

But happily they came through that first winter far better than I had any right to expect. Unpredictability is one of the charms of gardening, and for once it worked to our advantage. When I cautiously removed the leaves in March everything looked absolutely frightful and desiccated. But when I went out in April, trowel in hand, to dig up the dead I found that almost all the little wiry stems had greened up and were alive from base to tip. On the few that had suffered windkill the damage was not extensive and needed only trimming. I exchanged my trowel for hand clippers and spent a happy hour giving the plants a brisk haircut.

Heaths and heathers flower on new growth which comes in a sudden spurt once the weather warms up: trimming the lank stems induces the plants to spread outward, and it also enormously improves the appearance of any such planting. This is because heathers cling to their dead flowers for months, often all through the winter and into the subsequent spring. In unpruned plantings, the new flowers bloom on the tip ends of the stems above the tattered remnants of last year's bloom. If a heather bed is near the house and will be seen during the winter there is a great temptation to prune early in the fall and spruce up the planting. Old books often suggest this practice, and it is common custom in areas where the winters are less severe than ours. But around here I have a strong feeling that fall pruning weakens the foliage and makes it more susceptible to winter dieback, and for that reason I have always restrained myself. The excellent survival rate of our heather beds—for I later put in a second one near the main house—may be attributed to my failure to follow pruning advice. Or, of course, these plants may have survived out of pure Scottish obstinacy, just because I am only a poor Sassenach that cannot really ever understand them!

The Brides

There's a flowering plum in the form of a wide-spreading bush that turns into a fragrant cloud of white in late April and has never failed to give us a spectacular show from the very first season I put it in. Unfortunately, I don't remember where I bought it or even its name, although I have searched the encyclopedias to try to track it down. I have taken innumerable cuttings from it over the years, and these in turn have grown into striking bushes.

I enjoy the offspring, but they are never going to mean as

much to me as the mother plant, for that was in full flower on a bright afternoon when one of our daughters was married. The tent for the reception was stretched right over it, so the rather elusive scent was strong in the area where everyone danced. And fortunately no one was stung, although the sun was out, for the bees, which normally congregate around these flowers in hungry hivefuls, kept away, possibly scared off by the orchestra.

We took a considerable chance planning an outdoor reception so early in the year, and we cautiously installed heaters, but we didn't have to use them and the air was fresh and sweet. That particular daughter, after her wedding, was going to live permanently abroad in a tropical climate totally unlike the one in which she had grown up. For that reason she felt strongly about having a garden reception in spite of the possible risk of snow, and she got a good day with the plum, daffodils, and violets all in full bloom. And even though she has often revisited us, I don't think that she has ever again seen that bush in full bloom. By a sad coincidence it was also blossoming when we received the news of the premature death of her young husband, so it now holds the double symbolism for us of both joy and sorrow.

All our daughters had garden receptions, and I have strong flower associations in connection with them all. The scent of jasmine reminds me of the first of the family weddings, for we hired huge tubbed specimens to flank the receiving line, and their owner tucked a small sprig into the bride's bouquet. This I later rescued and turned into a successful cutting, and I now own a huge plant of my own and several smaller specimens all from that tiny twiglet. And the fact that they burst into bloom around that daughter's wedding date serves as a useful reminder for appropriate congratulations.

The third wedding was marked by an unexpected extravaganza of Oriental poppies that flowered at exactly that date. Before I knew the bride's color scheme I had worked out a delicate combination of white and pink peonies, pale pink sweet williams, pink and white Canterbury bells, and lots of spotted foxgloves for the flower beds which would go with anything. Since I had not expected the poppies to be out so soon, I had not taken their flaming colors into account. But although they wrecked the overall pastel effect of my original scheme, their strong colors fortunately blended rather than clashed with the bright shades of the bridesmaids' dresses, so I was able to pretend that everything had been carefully planned.

Obviously garden occasions are not possible for everyone. But it is wonderful what can be done in a tiny yard, and how much friendlier everything seems out in the open.

May

The Rites of Spring

I have just come indoors from carrying out what to me is the delightful annual ritual of walking around the garden, which is the suburban one, in which I have worked the longest, renewing acquaintance with familiar groups of plants and finding out how they have done during the winter rest.

In an old untidy bed on the back lawn, I search carefully for the red snouts of the peonies. These are the first plants I moved in this yard. We found them half-smothered in misery under a hedge and through luck, not knowledge, I transplanted them in August, the proper season. In consequence they have rewarded me with copious flowers ever since. Peonies loathe being buried too deeply, so one of my first jobs in the spring and the last before the snow flies is to scratch back the accumulated mulch and give these plants air space. Without that attention the curled red crosiers would be thin as they pushed above ground instead of thick and fat and pregnant with a minute bud. I also always look to see whether the shoots from the pots of lilies I forced for a daughter's wedding are above ground. This was one of my more magnificent muddles—I bought the wrong varieties and had wonderful flowers, so I am told, in the deserted garden months after she was wed. But the plants are still there and give my neighbors renewed pleasure and amusement each summer.

Next I check on the little bulbs that were set out in timid groups by my husband when we first moved in. He was new to gardening (he had always seen it done but never taken any active part), and he laid out this first planting with mathematical correctness. They are still his particular pride, and now that they are spreading in all directions the original unfortunate symmetry with which they were set out is vanishing, and each year the pools of blue seem deeper.

I hugely enjoy these tangible evidences of our years of pleasure with this garden, and I hope we shall continue with them for many more years. But there is also great pleasure to be gained from retrospective vivid memories of plants and places.

I have not seen the garden where I grew up for more than forty years, for my family moved away from it even before I married. But unless that garden has been radically redesigned, I can see through my inner eye where the water wagtails probably still nest near the pond. I used to inspect that nest each spring on my way to look at a larch which fascinated me with its fresh yellow-green needles and little cones like raspberries. I am sure I could still find red cowslips on a bank near the tennis court, for I can visualize exactly where it was they got a foothold against their rather pushy yellow cousins and staked out their territory.

And back in the corner of the orchard under a huge *Laurustinus* bush there should at this time of year be sheets of pale mauve Parma violets, rejects from the hotbeds, which had established themselves in that hidden, sheltered spot and brought swarms of bees winging in from the hives down by the wood.

These are happy memories, and someday when my memories

of this present garden must also be only in my inner eye they will be happy too.

Continuity

In the country garden we still possess an elm, a tall, lovely vase-shaped specimen of a fast-vanishing breed. We worry about that tree a lot: it is pruned and recabled each year, and we feed it on a regular schedule so that the ground after that treatment is encircled underneath by large green polka dots. We also spray it with great discretion, timing the sprays so that they will not harm orioles that may be nesting in the high branches. And the tree men and I are both exceedingly fussy about what is used and how.

The elm grows beside what appeared for many years to be a pile of field rocks covered with ivy. But during a recent bitter winter the old ivy completely succumbed, and after waiting more than a year for some sign of life at the roots, we eventually took on the huge job of carting all the dead away. To our astonishment, under half a century of ivy debris, we discovered the remains of a fountain with a carefully built stone wall surrounding it.

The fountain must have been much older than the tree, for it was incredible that anyone would ever deliberately have planted an elm where it was bound to ruin a rather elaborate piece of construction. And this fact suggested the possibility of a earlier garden, of which the fountain had been an important feature. Once we had that clue, it was possible, when the sun threw long evening shadows, to trace the faint outline of some kind of enclosure, possibly the foundation of old walls. When we realized the existence of this abandoned ground plan, we found a reason for the position of four extremely venerable junipers that are set out in the lawn. We had supposed them to be mere accidental features, but it now became obvious that they marked the four inner corners of the enclosure, as cypresses are used to this day in Italian gardens.

This must have been the very first garden to be constructed on that piece of land after it was reclaimed from salt marsh and cranberry bogs, and it must have belonged to a house of which there is no record. We ourselves razed a house on that particular plot of land about fifteen years ago, but the fountain and its enclosure did not belong to the house we pulled down, for it lies askew of the terrace that we retained when all the rest of the building was demolished.

Most of the ownership of this particular stretch of waterfront land is well known and carefully documented, and it was a considerable surprise to discover through this domestic horticultural archaeology that there had been an earlier unrecorded garden, and presumably a house, on it. It now seems likely that the building may have been a rather simple summer cottage—or what in those days were described as simple summer cottages, for the enclosed garden with a fountain does not speak of Arcadian simplicity—that was subsequently pulled down when the owners decided to build a larger, grander house on the same piece of land. And since both houses were put up by the same owners, the fact that there had been two had been forgotten.

When the second house was built, a very elaborate garden was laid out which was still in a reduced state of existence when I first saw the place. The old walled garden had been entirely swept away and replaced by a huge velvety lawn, the abandoned fountain, made picturesque by being covered with ivy, and a rose garden constructed alongside a grape arbor to one side of the lot. At some period when this reorganization was under way, a small wild elm must have sprung up beside the disused fountain, and since the change in ground plan probably took place at a time when the new house still looked rather large and without much comforting shrubbery around it, the tree was allowed to remain.

We know from old maps that there is an underground stream running parallel to the harbor at just this point, and obviously the little tree flourished mightily with a secure supply of sweet water always available and soon became the dramatic towering sight it remains to this day.

Unfortunately, in spite of all our care it now shows serious signs of stress, which is sad, for it has not really reached its prime; elms in good condition can live for well over a century, and we know that this land was not reclaimed as long ago as that. But we are fortunate to have it with us at all, for most of the neighboring elms have already succumbed to elm disease. There is now a new, faint hope that a cure may be on the way for this hitherto fatal ailment. But even if we can keep this killer at bay, hurricanes and other natural disasters have taken a heavy toll of the branches, an occasional premature leaf drop in late summer shows that all is not well with the root system, and each spring the litter of dead twigs on the ground suggests that winter die-back is increasing.

But still the old tree struggles on, and I hope it will outlast my time. And when it dies, unless there is man-made interference, nature will replace it with something else that will thrive with the help of that underground stream. I shall not see that replacement reach maturity, but I hope that it will give people the lasting pleasure we, and those who went before us, have had from those ancient junipers and the huge elm.

Tulip Time

The tulips did well this year; the late cool spring brought strong tough growth so the flowers lasted long and had a fine texture. And although there were a few very hot days during the two peak flowering periods, we did not have any of those devastating winds early in the month that are often such a feature of spring. This particular menace often appears impartially on both sides of the Atlantic just as the tulips start their show. My mother used crossly to call it "the tulip wind."

For years I had no luck with tulips after their first season, and not always even much of a show the first year because our changeable New England weather often ruined what had promised well. Indeed I got so discouraged with both the shortness and the trickiness of the period of bloom and the waste involved in the subsequent lack of success with the bulbs that I stopped growing them at all. But even allowing for the irritation of constant failure, spring did not seem like spring without some massed tulips, and I started using them again, although the second time around I handled them rather differently.

As a start, instead of buying brightly colored varieties I concentrated instead on shades of pale pink and white, for these do not bleach out so badly on our sudden blazing days and therefore give a longer show in fine condition. I also aim today for two peak blossoming periods, massing varieties that flower several weeks apart in different beds, and since I started this fail-safe plan I don't think we have ever been without one fine display.

For the beds nearest the house, which lose their snow covering first, I plant short-stemmed early tulips, using both the single and the double varieties. The warmth and shelter of the house bring these into flower very early, and violently hot weather is rare when they are in full bloom. These are also the best of the tulips for amateur pot forcing, and I always reserve a number of planted pots of the same type for the house. I have hardly any wind trouble with these early varieties outdoors, for they flower on a very short, sturdy stem, and the gales howl over rather than through them.

For the second show, which comes in flower beds around the patio, I use the middle-season Triumph hybrids which have exceptionally sturdy stems and seem almost wind-resistant. I plant these at least two weeks later than the early tulips so as to extend the difference between their flowering dates—a trick which sometimes works and sometimes makes no difference, depending on the condition of the ground when they go in: if it is already extremely cold, the difference in the planting time will have no effect. The enormous show of Triumph hybrids distracts the eye from the beds that held the early tulips, which by then are full only of green foliage. In sunny places in the area where bulbs are naturalized, and in warm pockets beside the garden door, I use little handfuls of species or botanical tulips, mainly *kaufmanniana, clusiana,* and *dasystemon.* These reappear reliably year after year, although I don't see much sign of any increase among them. But it is extremely pleasant to have some bulbs that you know will never let you down. I have given up the delicate lily and cottage tulips—I found they flopped too much—and the enormous box-shaped Darwin tulips are out of proportion in my small patio beds.

Another change in growing methods was in the aftercare. As soon as the flowers fade, I snap off the seed pods, leaving the long flower stems intact. And if you will force yourself to pick up all the scattered petals the beds will look perfectly tidy for a couple of weeks after the flowers are done. Then, when the second show is over, and while all the foliage of both types is still green, the bulbs should be carefully lifted with a garden fork—with a spade you are far more likely to slice through a bulb. I lift only a few at a time, for it is important to replant them quickly with the foliage intact and as much earth adhering to the roots as possible. We set ours in the suburban garden in sunny open places among the naturalized bulbs about five inches below the soil surface with a lot of fresh compost underneath them. Once replanted, the area around them is heavily sprinkled with an organic fertilizer, either Milorganite or Bovung, which is then watered in.

The tulips remain gaunt and upright while the foliage slowly dies, but during this process the leaves and flower stem send back sufficient nourishment to the bulb to plump it up and reset an embryonic bud. If you make the mistake of resetting the bulb too shallowly, it will split during this final effort, and the small bulbs that look like the offsets of garlic will not set bud.

I have no success if I follow the practice of allowing the bulbs to ripen in a warm place above ground on drying racks. My bulbs shrivel away to nothing under this treatment. Only by allowing that miserable foliage to die disgustingly have I been able to achieve subsequent success with tulips.

For that reason I control my itching hand which longs to pull these miseries out. Instead I let them remain like bleached bones in a desert, the ghosts of the plants that have been and the token of flowers to be.

Easy and Attractive

A delightful, slightly unusual flower bed can be produced in a sunny place by massing petunias and fronting them down with an edging of curly parsley. This is a trouble-free combination once the actual work of the planting is over, for there is no upkeep except cutting the dead heads off the petunias and clipping parsley for the kitchen. And the plants can be set to grow so thickly that weeds will not take over. If you are ambitious, the effect can be made even more striking by using parsley rather like an old-fashioned knot garden in ribbon lines within the border as well as along the front.

The job of getting the planting area ready and sowing the parsley seed should be done early, as soon as the ground is workable, well before the petunias are on sale, to give the parsley time to sprout and show the outline of the planting plan. For best results the ground should be dug. You don't need to be heroic about this and try to reach China; just turn the soil over to a depth of about six inches. This will uproot the weeds so the land can be properly cleaned off and will loosen compacted earth sufficiently to let in the oxygen necessary for good growth. After digging, spread a thick layer of dry steer or cow manure on top of what will be very rough, lumpy-looking ground, and rake this in thoroughly. Raking smooths the soil surface and gathers together innumerable stones that mysteriously will have risen to the surface, while incorporating the manure into the ground. Follow this up by soaking the newly dug bed, not just a light springling but a thorough drenching, and then leave the ground to settle for a couple of days. Slow natural settling of newly dug soil is important in preparing land for planting.

To sow the parsley seed, make furrows with the edge or handle of the rake. A seed bed must always have a firm base for good germination, so either tamp down the inside of the furrows with the back of the rake or walk along them heel to toe. Next strew the seed along the flattened interior of the furrows; since you mainly need this for exhibition, the seeds should be close together though not on top of each other.

The next step is characterized by the experts as folklore, but it has worked like a charm for me for years and I have always regretted it when I have not followed the old ways. Once the seed is down in the furrows, take big kettles of water that have come to a rolling boil on the stove and walk along the furrows pouring the boiling water over the exposed seed. After that pull half an inch of dry soil over the steaming furrows and stamp that down. Parsley left to itself germinates slowly and extremely unevenly. Often after waiting for weeks, it comes up in little clumps with long empty stretches between. This is not what you want with parsley used as an edging, and though I am assured that scientifically there is nothing to the boiling water treatment, if you will but try it, the parsley will appear sooner and without those empty spaces!

Once the little plants are up and you can see the outline of your edging pattern, the time has come to set out the petunias. A petunia bed looks vastly more effective if it is planted in blocks of solid colors and even more delightful if you shade the colors. Unless you are planting inside dividers of parsley, plant the petunias eight inches apart. If the plan is to have rows of petunias behind a solid border of parsley, make the rows a foot apart and each individual plant eight inches apart. You can easily work out the number of plants you will need by poking holes with a stick according to these rough measurements.

For the color scheme, a mass of white in the center looks nice, and it can be flanked on each side with blocks of a pale pink color, which in turn can have plants of a darker shade of pink running from them to the outer limits of the border. Another effective design is a block of dark blue flowers in the center, pale blue on each side of the dark flowers, and white running to the ends. And never forget that the larger the blocks of color the more striking the effect. A bed of mixed colors will not be nearly so interesting visually, though it will grow equally well.

One of the problems is getting the right varieties and colors in the petunias. Ideally we should grow our own, using the new single multiflora F1 hybrids that flower at the same height. But unless you have a greenhouse with plenty of space, this is impossible for the home gardener, and you will have to buy the petunias. Most garden centers sell annual plants either in containers a little like miniature egg cartons or with the plants individually planted in peat pots. Both make for very little root disturbance when the plants are set out, but the peat pots should be well soaked in warm water before planting. If the petunias come in a single container with no divisions, knock the whole block still intact out of the flat and cut the plants apart with a sharp knife. After planting, always firm the plants in individually with your hands. It will get them off to an excellent start if you follow up the planting by watering each individual petunia with water-soluble fertilizer in a weaker dilution than that suggested by the maker.

Parsley is a biennial: it will not flower the first season but will make a thick edging. The second spring it will reappear more flourishing than ever and throw a lot of flat white flowers rather like Queen Anne's lace to which it is related. Your parsley edging can be reused a second season if you sternly cut off the flower stalks as they form. After the soil warms up, hundreds of little petunia seeds will also reappear a second year, for you never succeed in cutting off every petunia seed pod no matter how faithfully you try. You will not have the same excellent show if you try to make do with these self-seeded plants. They may, it is true, turn into larger, lusher plants than those of the previous year, since they have developed untouched from seed. But the colors will all be wishy-washy and mainly a rather dull pale purple, for the first descendants of hybrid seeds invariably revert to a rather uninteresting ancestor. So buy new petunias and reuse your parsley and get two years' pleasure out of one year's work!

Nettles

One of the pleasures of writing a newspaper column is the quantity of peripheral information that subsequently is sent to me by kind readers. Some time back, for instance, I mentioned in an article the part played by the alien European invader the stinging nettle (*Urtica dioica*), its name all too accurately derived from the Latin word *ure,* to burn, in redeeming land defiled by man.

This comment produced a great deal of correspondence, all surprisingly in favor of what to me has always been a plant of rather questionable attraction. Nettles are the great bane of an English childhood. They line riverbanks and hedgerows and sting the ankles of unwary passers-by. They are or were also used by farmers as a form of watchdog. Big stands of a particularly ferocious variety known as the Roman nettle (*U. pilulifera*) were planted deliberately by a local farmer all around his cherry orchards to deter marauders such as myself who yearned after the ripe fruit. And those stands served the purpose admirably, for it would be a very brave child who waded through them. Nettle stings do not have the long-lasting effects of poison ivy, but they can be extremely painful for several hours after contact.

But even the ordinary perennial stinging nettle was something to be avoided, and it also was a rampant grower that was hard to control. My father, wielding a scythe like Father Time, used to cut down a big stand that sprang up each year in a new place beside the chicken run three or four times during the growing season, for the only way to eliminate nettles is by regular, constant cutting.

The attitude toward them in this country seems to be much more benevolent, possibly because although they have naturalized themselves in the Northeast they are still not very obtrusive. One of my correspondents, an English exile like myself, so deplores the local dearth of nettles in her area that she sets aside a special place in her yard where she cultivates them. And when she moves, which appears to happen quite often (and for which burden, as a fellow horticulturist, I offer my most profound sympathy), she takes transplants of her domesticated nettles along with her to establish a new patch. In her kindness she supplied me with the name of a dealer where I too can buy nettle transplants if the urge so takes me, but I already have so many other problem plants in my yards that I don't think I am going to take advantage of this piece of information. Also, since American children in this locality are not trained to avoid nettles, there might be some harrowing scenes with the grandchildren if I were to introduce them into the garden!

But my rather lukewarm attitude toward stinging nettles as a horticultural feature does not detract from my appreciation of their usefulness in recovering ravaged areas of land. Nor do I want to include the harmless Dead nettle (*Lamium album*) in this cold shouldering. I find this white, spreading spring flower a pleasant sight in an open border, and although it is not reliably hardy in our yards, I try to keep the flower beds supplied with plenty of it.

To judge from the letters I received, there are far more modern uses of stinging nettles available to us than I had realized. If the stems and leaves are steeped in water for three weeks they make, I am told, a highly potent brew which when strained and sprayed over growing plants acts as a stimulating foliar feed. I get the impression that, like another homemade fertilizer, crushed eggshells in cold water, nettle fermentation should take place at some distance from the house and in a tightly closed pail; the process, it seems, is highly pungent.

Another writer uses nettles to reclaim land in poor condition by laying them down thickly as a mulch which is subsequently turned under. I do wonder whether every nettle is grasped firmly before it is put down, or whether the mulcher has a heaven-sent immunity to the stings from the fine leaf hairs. I realize that dock leaves, which always grow in the same neighborhood as nettles, clapped upon an injured area dilute the pain. I have pressed many a dock leaf into service for that purpose myself—but mulching with nettles would seem to me to call for acres of docks, and that in turn bespeaks poor land!

Nettles are also good for those who suffer from an iron deficiency. The young tips can be cooked alone or combined with spinach in a soufflé. My slightly cynical correspondent on this matter assured me that the dish was always an immense success just as long as the basic ingredient was unknown.

I have always been vaguely aware that nettles were supposed by our forebears to be of use in cases in which people were affected with gravel and stones in the urinary tract. In her delightful book *Early American Gardens,* Anne Leighton mentions a great many other ailments for which nettles were supposed to be the appropriate cure.

Up to now I had believed in the rather austere verdict on stinging nettles offered by the Royal Horticultural Society's *Dictionary of Gardening,* which states: "None are of any value." But now it seems that I should update myself on this matter and try to be more adventurous.

Trials and Tribulations

So far there has not been a single restful moment in the country garden. The odds seem to be against us, and the preliminary work to be done immediately appears more demanding than usual. In the flower garden most of my biennials succumbed to the winter cold, although there were a few pinks and Canterbury bells to greet us with a fine promise of potential bloom. But their moment of triumph never arrived, for dogs, rolling in the flower beds in the early hours of the morning, smashed practically everything down. I can't say that I entirely blame the dogs; they were attracted to that area by a pungent barnyard smell coming from a mulch of lawn clippings that we had laid down. To make no bones about it, the garden smelled, and that lured the dogs.

Why grass clippings, which normally have a faint hay scent,

should have turned so rank I do not know. Maybe the unending wet weekends which we have suffered late this spring after the grass started growing had something to do with it. But even before the dogs got in, we had decided that something had to be done about the pigsty atmosphere and had started to cover the offending mulch with piles of half-rotted leaves. These freed us from the smell, but set up new difficulties of their own. For although the leaves were nice and moist and clinging closely together when they were put down, the continuous fierce hot winds which have been another unusual feature of this early summer dried them out very fast. As a result the leaves blew about in a premature autumnal scene that smothered many of the small annual seedlings that were just emerging. So the leaves in turn had to be held down with a fresh layer of newly cut grass, which this time was carefully dried out in advance, and that finally solved the problem—after a great deal of time and effort had been wasted.

Oddly enough, while I was battling with this particular difficulty I got an urgent letter from a TV viewer asking for help with her compost pile. She had recently moved into a new house and had started a pile composed entirely so far of grass clippings and had run into exactly the same olfactory problem as us, but with the added difficulty that the neighbors were complaining. The only person who might complain about ours was myself, for the area where our troubles were poisoning the air lies beneath my bedroom window, and I had found it necessary to keep the window on that side of the room shut, so I knew what my correspondent's neighbors meant. Her problem, however, had a fairly easy solution. When piled-up grass clippings smell, which can happen in hot steamy weather with a big pile, the difficulty can be controlled by covering the heap with a thick layer of earth. This eliminates any cause of complaint and adds nutriments to the pile. But I was not able to follow my own advice in our flower garden, because I was troubled not by a heap of piled-up grass but by a rank smell coming from layers spread between the rows of flowers. And although putting down earth would have solved the smell, putting earth over a juicy layer of decaying grass would have resulted in a bumper crop of enormous weeds—something that mulch is supposed to discourage, not invite!

I use all kinds of mulching material for weed suppressing, both inanimate and organic as well as living plants. About the only thing I do not use is newspaper because it looks dreadful. But since there is now a lively controversy raging about whether the chemically composed printers' ink in the paper will contaminate the soil, perhaps it is as well that I never started newspaper mulching—although I have often been urged to do so by conservationists who deplore the short use we make of paper pulp, which unfortunately is largely obtained by felling trees.

Although mulching the flower and vegetable gardens is one of the first and most important actions we take after we move to the country garden, I grudged the time it took. I had another very urgent job awaiting attention: to get after the overexuberant spread of the ferns which I use as ground-cover plants and which had also run wild in the wet spring.

Ferns are some of my favorite weed suppressors, but they are highly aggressive takeover artists in both sunny and shady positions when they are happy. Ours are extremely happy, and this looks like a bumper year for ferns: they are scrambling out of their allotted areas in a manner that demands fast action. Already their canopy is green and lush, and to a bee flying over them they must look like a tropical rain forest.

Ferns are some of the oldest plants on earth; they antedate almost every other growing thing except the algae, lichens, and mosses. But the tenacity that has kept them successfully alive on this planet for unnumbered years means that once established they are hard to eradicate. Fond of them as I am, it is only fair to warn you to think carefully before you decide upon fern ground covers—they may become more than you had reckoned on. If ferns get ahead of you, you can try to control them by digging out the spreading clumps. This I find very hard work and not entirely successful, for ferns spread by underground root stocks called rhizomes. You may dig out a clump, but the rhizome that produced it from a distant plant will still be there; once your back is turned, it will throw up a replacement. In my experience the simplest way to keep ferns in their place is thoroughly to rake over the area where you do not want them with a deep iron-tined rake. Raking will damage and even uproot the creeping rhizomes, thus putting an end to the insidious spread. It should be done in the early summer and again in the fall near any fern planting that you wish to control.

Digging and transplanting ferns from the wild is a chancy business, for almost all wildlings resent being moved. Furthermore, many ferns are on the conservation list, that is, the list of plants that are endangered because of building or human greediness. Unless you are a fern expert it is difficult to know what you are about, and it is far better not to take any chances. Instead buy ferns from one of the many nurseries that specialize in wild flowers, for they will be able not only to provide you with the varieties that suit the areas where you are planning to plant, but also give you local cultural advice. This, of course, varies according to the soil conditions, but in general ferns do best in slightly acid soil and with rich humusy soil.

I bought a mixed group of shade-loving ferns many years ago —I already had plenty that like sunny places—and I can honestly say that the various locations where both types have been planted have never since given me any trouble except for the necessity of establishing some kind of control. There are not even any weeds; the ferns overwhelm everything. This admittedly does mean that I also never see much more than a glimpse of the trilliums and autumn crocuses that were set out in some of the same areas, and that even some tough midcentury hybrid lilies also planted among the ferns are having to fight for their continued existence. The fern growth is so exuberant that very little else is able to survive among them. As a general warning, therefore, be very cautious about combining ferns with other plants—the chances are excellent that the ferns will be the unchallenged victors.

But there are a few members of the family that are not quite such takeover artists. These include the various osmunda ferns such as the cinnamon fern (*Osmunda cinnamomea*), the interrupted fern (*O. claytoniana*), and the superb royal fern (*O. regalis*), which all grow in tall stately clumps and are not nearly so aggressive. Indeed I get the impression that in the family circle, osmunda ferns would prefer to disassociate themselves from most of their more pushy relatives. These varieties, which look particularly lovely as they unfold, can be effectively combined in shady areas with the billowing forms of the many hostas to produce delightful contrasts in texture and height.

All my royal ferns are divisions from a twitch I stole years ago from a rented garden, and they have turned into an extremely worthwhile horticultural theft. In my own defense I must add that the house and garden where they originated was slated for development, and the magnificent clumps that grew there have now all vanished—while their descendants give enormous pleasure in both our yards.

But their spreading cousins are not going to please me much longer if they get out of hand; now that the mulch problem is at last settled I must get to my raking. Let us hope that nothing else goes awry in this odd year, and that from henceforth this season I can garden according to a normal schedule.

Try Some Orchid Cacti

This has been an exceptionally fine year for flowers on the orchid cactus, that tree-perching plant from the tropical forests of the New World that used to be known as *Phylocactus* but is now listed as *Epiphyllum.* This plant entered my life as an indoor gardener relatively late when, some twenty years ago, I read an article about it in the gardening section of the *New York Times* and was so entranced by the description and the illustrations that I decided I must have one.

This did not prove to be all that simple, for I found it hard to get hold of a plant. I went, article in hand, to several sophisticated florists before I found one that undertook to order me a specimen. Here in the East, we were far slower than indoor gardeners on the West Coast to take advantage of the potentials of this epiphyte cactus for our plant windows. And to this day, even though it is now easy to buy specimens locally, the catalogs of West Coast nurserymen who specialize in these plants and their near relatives make my mouth water.

My first plant was the now familiar red *Epiphyllum ackermannii,* but that first plant was a spindly little misery that I don't think ever flowered well. It still sticks irritatingly in my memory that this plant I had got with so much difficulty arrived in a broken pot because the deliveryman had dropped it. So I was immediately faced with the problem of repotting something whose particular needs I did not understand. I had no firsthand knowledge of how to handle orchid cacti, for at that time I had never seen them in the wild and obviously I had mislaid that original article. I made the fairly reasonable mistake of supposing that the best way to handle the invalid was to treat it like a ground cactus from a dry climate and give it sandy soil and very little water. By so doing, I nearly starved my poor misery to death.

But orchid cacti are exceptionally tough plants that recover fast from mishandling. When I realized that I was getting nowhere with my purchase and took the trouble to look up its culture—which involves deep crocking, very rich soil, and lots of water—in the library of our excellent Horticultural Society I had no further difficulties. My plant soon turned into an excellent foliage specimen, but the flowers were scanty and often the buds failed to open, dropping in exactly the same irritating fashion as those of its near relative the Christmas cactus. Eventually, however, through trial and error (and partly I suspect through exasperated neglect) I discovered that the secret to lavish, successful spring bloom is to keep the pots of orchid cacti almost completely waterless from late October until early April. And furthermore, they do best if this unnatural treatment is meted out to them under quite cold conditions so that they fall almost entirely dormant. Quite what natural condition in a hot, humid rain forest approximates this handling I have not worked out, but the result of this hard-hearted handling here in New England are buds in almost every notch of the modified stems a few weeks after lavish rewatering begins in the spring. So I now enjoy a collection that includes delicate shades of pink flowers, others with purple stripes, and many of soft creams and yellows as well as innumerable offspring of the first flaming red specimen.

I keep them all in rather small pots with an unusually deep layer of internal drainage. Sometimes I enlarge the bottom drainage vent, which is easily done with an ice pick and a light hammer as long as you start breaking very small nibbles around the edge of the original aperture. If you prefer to use plastic pots—and I have grown excellent epiphyllums in plastic—put in at least three inches of broken-up crockery as internal drainage after enlarging the vent holes with a red-hot screwdriver—a very smelly job but perfectly practical. We use pure compost laced with sand for the potting medium (perlite can take the place of sand), and we don't pot too tightly. The one cultural essential is that surplus water can rush away through the roots, and hard potting can prevent this fast runoff. The books say that only the new growth carries spines, but I find this hard to believe. Repotting is a dreadfully thorny business. Those long flat blades may not look very menacing, indeed you may not even notice the thorns, but the plants are in fact a terror to handle, and you will feel the aftereffects in every finger! I shift my plants as little as possible but not just for the sake of my personal comfort; in my experience the plants flower better when potbound.

It sometimes happens that the technically modified stems that constitute the foliage grow so long that it is impossible to keep the small pots upright. I solve this problem by sinking the planted pot inside a slightly larger one and filling the interstices with stones. This provides the necessary steady base without unending cutting back or repotting. The blades can be easily pruned, however, if that suits your taste better, and the time to do this is immediately after flowering. Don't trim them all at once or there will be no flowers the following season: the buds form on old wood.

Slugs and snails eat holes in the blades, but an injured stem will survive, looking perfectly frightful for long enough to enable the plant to throw up new growth, after which the old stem should be cut out. In the summer, the potted plants enjoy a warm, sheltered place where there is at least half a day of full sun. If you hook plant hangers over their pots, they are at their happiest swinging from trees. If this is impossible, set shallow saucers of beer among the pots to lure the slugs to drunken death, for slugs will take beer in preference to the juiciest leaf. During hot weather water epiphyllums daily, and I use a fish emulsion fertilizer once every three weeks. In late October when they have to go back to town, our plants are rolled in newspaper to guard the new growth from injury, for the blades get cut very easily if they are tied back. The pots are then set on the shaded north side of the greenhouse almost touching the glass where in midwinter the night temperature falls into the low forties.

For the next four months I do absolutely nothing about watering them; they get only the drips that run out of the plants staged above them; before I had a big greenhouse, I used to give orchid cacti this waterless winter rest in a north-facing window of an icy attic where there was no heat!

Early in April heavy watering should be started up again, and to speed bud production the pots should be moved to a warmer position. At this point many greenhouse growers run into a problem with the orchid cactus. Unless you go in for regular preventive spraying, which I do not, most greenhouses contain hidden away among the plants a few stray aphids which normally are entirely controllable by hard hosing with cold water. But the moment the epiphyllum buds begin to elongate, the hitherto unimportant aphid problems will suddenly explode into a menace. The pests collect all over the swelling buds so thickly that they look almost like a green scale. As far as I can see, these infestations, deplorable as they look, do not harm the epiphyllum buds. But burgeoning colonies of aphids are a nuisance in a greenhouse, for invariably the multiplicity of their sticky population explosion spreads to the other greenhouse plants. To avoid this problem, when the time comes to move the orchid cactus into a warmer position I go over each plant carefully to make sure that it is free of pests, and then move the entire collection upstairs into a cool empty window with plenty of morning sun where they can neither infect others nor become infested themselves. As soon as the first flowers open, the amount of water the plants receive is greatly reduced, and the pots come downstairs where the flowers can be enjoyed. In mild seasons it is sometimes possible to stage the epiphyllums outdoors by mid-May; if the weather is still too cool I make a special show of them in a plant window. Epiphyllums do not combine well with other plants, but they are superb massed together, and a single specimen can look exciting in a jardiniere. Each individual flower remains open for only a few days, but the bud set with this treatment is usually so lavish that the display period is very extended—and even indoors almost all the buds will open.

Epiphyllums in bloom look rather fragile; the buds on long arched necks seem doomed to be knocked off. But in fact they are unexpectedly tough—we have often moved heavily budded plants crammed into cartons without any loss of bloom. And happily the plants once in flower need very little attention; heavy watering will cause bud drop, and a little deliberate neglect suits them far better.

If our plants are still full of buds when we move to the country, we restage them immediately on the covered porch that faces the sea. Here it is noticeable that the final flowering comes in stronger colors and lasts even better in the moist air off the sea. These are ungainly tiresome plants to carry during their inoperative months, for they cannot by any stretch of the imagination be treated as decorative foliage plants. But the long-drawn-out spectacular flowering period makes the waiting well worthwhile—and you may even get a fall dividend of a few scattered autumnal buds if you are kind to these ungainly ugly ducklings!

June

Summer House Plants

People who enjoy house plants but have no garden often feel a little out of it and wistful after the weather starts to warm up and all the hearty outdoor types are busy planting or carrying their potted plants into the yard for a summer holiday. And, undoubtedly, with the notable exception of African violets, most house plants will benefit from being outside for two or three months—but these benefits will come only if some rather strict ground rules are followed.

No matter how sun-loving a potted plant may be, or how bright the window in which it has been kept all winter, every house plant must start off its sojourn outside in deep shade, for immediate outdoor sun will inevitably scorch and disfigure the leaves. Light coming through glass bears no relationship to direct light, and even greenhouse plants at first need protection from full sunlight. Acclimatization to outdoor light must take place slowly and by stages. From a week in deep shade under a bush, potted plants can go into dappled shade under a high, canopied tree, and then after another week of this shifting light, the sun lovers can go into full morning sun. If possible, never keep potted plants in an exposure that faces the blindingly hot afternoon sun. This is almost too hot for everything in pots, and calls for morning and evening attention to watering, which is a bore.

Indoor plants, even when outside, should be kept confined in their pots. It is a great temptation to knock them out and plant them directly into the soil, for this saves the need for daily watering. But plants that are taken out of their pots spread their roots in all directions, and it is almost impossible to cram these back again when the plant has to return indoors in the fall. If you want to save yourself trouble with their care, you can make the extra effort of digging a trench and laying a plank along the bottom of it. Standing the pots on the plank prevents roots from wandering through the drainage vents into the surrounding soil, which leads to injury to the plant when the pots eventually are lifted. A plank also prevents worms from entering the pots. But it is only fair to warn you that digging the trench is a considerable nuisance, and since your house plants are not all in the same size pots, you will have to do a certain amount of standing the small ones on other pots, which in turn rest on the plank, in order to get everything level before you throw the excavated earth back around the buried plants. Personally, I would rather water regularly than let myself in for such a job, but then I don't go away much in the summer and am around to take care of my house plants. For those who take long annual vacations, or even long boating weekends, the trench burial of treasured plants may be worth the work. House plants summering outdoors have to be given as much regular attention as they had indoors; this means not only regular watering, but also inspection for pests, pruning, and the elimination of weeds from the pot soil. Summer with all its diversionary delights can make all this care impossible, and house plants that have been left to their own devices during this period of rest and recreation often come back inside looking worse than when they were put out!

By contrast, the plants of gardenless gardeners receive continuous care, and with a slight change of handling to make allowance for the difference in the weather, they often come through the hot months in better shape than those that went outside so triumphantly in the spring. One important change is putting house plants in a different position. In winter the best place for all plants, sun lovers or foliage specimens, is a sunny window. In summer, any window with full sun is probably far too hot for year-round house plants, and these will do far better if they can be moved to a north or east exposure. If this is not possible, pull them back from the main window so that the sun does not fall on them. If the only possible place is near an air conditioner, take care that the plants are not in the direct line of a blast of cold air; they will enjoy a cool atmosphere but not a howling draft.

Potted plants should be watered only when the surface soil looks and feels dry to the touch, but during hot weather in non-air-conditioned rooms, house plants need a lot of extra misting to make up for the moisture the heat pulls out by evaporation through the leaves. Air pollution being what it is, plants in a room with a continuously open window should have their leaves washed once a month, more often if you spot dirt accumulating. Smaller plants enjoy a weekly shower in the kitchen sink. But since hot, humid weather produces disfiguring mildew on damp leaves, indoors and outside, misting and washing off should be done early in the day to allow the foliage to dry completely.

To all house plants summer brings a spurt of growth which can be handled three ways. If you like the size of the plant and its container size fits your room, you can keep it properly proportioned by top and root pruning. Knock the plant out of the pot and with a sharp knife pare about an inch off the complete circle of the root ball. I find standing the plant upright on a hard surface and slicing downward does the best job. After cleaning out the interior of the old pot, reset the trimmed root ball with plenty of fresh soil pressed in firmly around the open edges of the container, and balance the root pruning by cutting back a lot of top growth as well. If you want your plant to grow bigger but do not want to increase the size of the container, early summer is the time to fertilize to stimulate top growth. Water-soluble fertilizers are best indoors, but follow the directions exactly—a little more than a little is by much too much! Stop feeding by midsummer so that the new growth has time to toughen up. If you yearn to have an enormous tubbed plant indoors, early summer is also a good time to transfer a strongly growing plant to a much larger container. But go very easy on the watering if there is much more soil than root ball. Until new roots are actively at work spreading out into the extra space, the earth in the large container can easily become sodden with overwatering—and that's the route to rot. All these operations call for a considerable expenditure of energy on the part of the plant, and this energy is easier for them to come by when there are long daylight hours.

A certain number of more temporary flowering plants can be used to perk up that sunny window that is too hot for foliage plants. If the sun streams in all day, geraniums may thrive. Don't try the ivy-leaf or hanging types—these will turn yellow—use the familiar zonals that also do well in sunny windowboxes. Better still on sunny windowsills are the innumerable small or even miniature geraniums. These are often marketed in the peat pots in which they were struck as cuttings. Before you repot a specimen in a peat pot, break off every scrap of pot that rises above the soil surface of the plant inside it, and for even better later growth, soak the plant in its peat pot in warm water for a couple of hours and then carefully peel as much of the pot sides off the roots as possible, unless these have already penetrated through it. The reason for this precaution is that any peat product left in the open dries out to cardboard consistency and if the tops of the pots project above the surface of the new container, they will wick out the water in the soil, thus stealing it from the roots. Furthermore, if the peat pot sides were hard before the geranium was repotted, they will not soften up no matter how much water is poured into the container, the roots will never break out of their strait jacket, and the plant cannot thrive. Too many people unwittingly pave the way to a lingering death for their plants by putting them straight into the ground or pots still in those hardened peat pots. Disintegrating peat pots, which sometimes madden us when we buy them, are in fact a much better sign for the ultimate health of the plant they are supposed to contain than those with stiff, unyielding sides.

For a bright, sunless window try various begonias; they all do well inside during the summer with the possible exception of the tuberous begonia, which needs more air. The familiar wax or ever-blooming variety should have some of the long stems pinched back every week to prevent the plant from getting leggy. Impatiens or busy Lizzie also flowers well in a bright window, but with both these plants try to get double varieties, for they need less picking up after. Ferns, which are tiresome indoors during the winter, also thrive at an open sunless window, and if you can get some of the fussier types—the maidenhairs, for example—established and accustomed to your ways during the hot months, you may be able to continue to grow them successfully when summer is done. Those impossible gardenias are also excellent summer-flowering plants for a bright sunless window; too much sun will yellow their leaves, but they will accept it in the morning. Gardenias and ferns must always have a very deep layer of pebbles in the saucers underneath them from which surplus water can evaporate at all times. All the many echeverias are also willing visitors to a summer windowsill.

In fact, be a little more adventurous at this season; take

rather different care of your established house plants and try experimenting with inexpensive summer plants. You pay your money, and you take your choice—and fortunately for those of us who must stay at home, the choice can be very large.

Regal Geraniums

This is going to go down as the year in which I muddled the regal geraniums, for my plants have been so poor that I have not dared move them to the window.

Pelargonium domesticum, otherwise known as the show, fancy, decorative, regal, or Martha Washington geranium, seems to be making a comeback as a plant for amateurs to grow in a greenhouse or cool bright plant window. This is all to the good, for regal geraniums are delightful plants that produce a huge flush of bloom with spectacular effect. They were enormously popular in the last century, and the descriptions in old books of the colors and varieties then available are mouthwatering. But they fell out of fashion, possibly because of their susceptibility to various ailments, and almost vanished from the scene. Now, however, the species is being worked upon by the West Coast hybridizers and various European specialists who have managed to breed greater resistance into various strains. And though some of the past beautiful colors seem to have vanished forever, there are still many enchanting soft pastel shades available that make this a plant which fits in everywhere and with everything else.

When I started indoor gardening no regal geraniums were around; in fact, I had never consciously laid eye on one or seen it grown. My first plant came from a roadside stand on a back road along which we then had to travel to reach the country garden. I noticed a totally unknown plant on sale beside a farmhouse and pulled up to see what it was. It was a tiny, purple, pansy-faced geranium, a very inferior version of this particular geranium family which, to this day, is what you will get if you ask for a Martha Washington geranium without specifying that you mean the large flowering type. And it was on this poor little plant that I learned by trial and error, and through consultations with old gardening books, the small tricks of managing the species. For with the decline of interest in the plant, a proper understanding of how to handle it also vanished from public knowledge. And even now that there are more around, I wonder as I watch one being carried from a shop whether the purchaser knows how to get the best out of what has been bought, for the average salesperson rarely can provide correct information.

For although these are geraniums (correctly, pelargoniums) and therefore close relatives of a plant that almost everyone grows, they are not outdoor plants. Hot sun will fade the flowers, and moisture, rain, dew, or your careless watering will rot the petals. The main period of bloom comes early, in mid-May, and the regals will not continue to flower for the rest of the summer like their cousins the zonals and ivy-leaf geraniums. Once the first huge flush of bloom is over—which takes time, for properly handled, every shoot of foliage carries a bloom—some of the older varieties set no more bud for the rest of the year. Other, newer hybrids produce occasional flowers until later in the season but never in the same quantity as that first wild careless rapture.

Regal geraniums are also much messier plants than their cousins. As the petals die, they don't wither away discreetly, still attached to the flower stem. Instead they shatter and fall all over the windowsill; so this is a plant that needs to have a dustpan in constant attendance. To get later flowers, once the main show is finally over, cut off the dead flowers at the stalk but do not cut the foliage itself. Put the pots in a bright place where they are protected from rain but don't have too much sun—a summer shaded greenhouse or a bright open window is excellent—and feed with weak liquid fertilizer every second week. If you have some of the newer hybrids, a few shy flowers will eventually appear on the tip ends of the side shoots until early August when new growth stops. Then move the plants outdoors into a sunny place where the foliage will be ripened. Keep the pots well watered but cut out the fertilizer. After Labor Day the plants should be cut back, though not all the new growth of the current season should be taken off. Simultaneously they should be root-pruned and then repotted in rich soil which is rammed firmly into place. The new pot should be a size smaller than the one in which the plant bloomed. If a lot of strong new growth has been cut off, this can be used as cuttings to produce fresh stock. It is a good plan to keep the stock built up by taking a few fresh cuttings every year, for in time old plants diminish in vigor. The method is simple. When new growth is ripe enough so that it snaps when bent over, it should be cut in four-inch lengths and then trimmed straight across with a razor blade where a leaf meets the stem. The material then is stood one-half inch deep in boxes containing a 50–50 mixture of clean sand and damp peat moss. The medium should be kept moist but not sodden, and wilting prevented by a light misting of the foliage. Roots will form in about three weeks. Cuttings can also be struck under plastic tents, but treated this way they are more liable to mold before the roots are fully developed, or else get a disease called "blackleg" which is a clear enough term to need no description!

After the old plants have been cut back and potted down, they will react to this rather rough treatment by falling into a condition of near-dormancy, and during this period the trimmed roots will slowly heal. But, unlike zonal geraniums, regals cannot be forgotten and left almost dry during this resting time, which comes during the early months of the winter. They must have light, a cool position, and light but regular watering. My pots are set on a high shelf close to the glass of the greenhouse, a place where it is extremely hard to give anything too much water, for it all pours down my sleeve as I stretch upward. This appears to suit the plants! When I had no greenhouse, I wintered regal geraniums in a bright, very cold attic window. When the days start to lengthen in late January, a period when to us the worst of winter is just starting but which to plants is true spring, fresh green growth will appear along the trimmed stems. At that stage move them to a place where they can have better, fuller watering. When the foliage really starts to take off, move the plants up into their flowering pots, which can be two sizes larger. Again always tamp the pot soil down firmly. Do not pinch out the ends of this new foliage growth; old plants will bush out naturally as they develop, and the flower buds set on the terminal point of this new growth. This essential piece of information is not made clear enough to would-be growers of this plant, and not always to the florists who market them. I have been offered plants that had been so thoroughly pinched out that, although they were nice and bushy, there was not a bud in sight nor any chance of one appearing. Yet the salesman blandly informed me that plenty would soon appear! Never buy a regal geranium without examining the foliage ends very carefully. Unless you can see minute buds forming don't buy it; they may, mistakenly, all have been pinched out. It is in this matter that the regal geranium is so totally different from the zonal and ivy-leafed species which do bush out after tip pinching and still flower profusely. Big regal geraniums, no matter how cool the place in which they are raised, have a tendency to be weak-stemmed, so when the buds are well developed, staking with green wool and thin bamboo sticks is in order. If you don't stake, the flower heads are inclined to be top-heavy.

The cuttings that were fall-rooted have to be treated differently, for these cannot take the very dry resting period the older plants demand. Instead, carry the cuttings on in three-inch pots with regular watering in the coolest brightest place you own. In early January, if growth is good, pinch out the tip. This sounds like heresy, considering what I have just said about pinching, but unless you give a single stemmed cutting a light pinch it will not branch and you will grow yourself a standard geranium! After that first pinch, side shoots will develop. These you can also tip-pinch when they are two to three inches long—if the date is no later than mid-February. With this treatment you will not destroy buds—they are not yet forming—and the plant will bush out very satisfactorily. A cutting that has not rested has more growing leeway than an older plant. A cutting can be moved up out of the three-inch pot into something considerably larger once the foliage has started to bush out.

Like everyone else who grows these plants, I am troubled by whitefly. During the summer regular hosing off with a hard stream of cold water, with particular attention paid to the undersides of the leaves, usually holds these pests in check. The material that I expect to use for cuttings is always held down with a stone in a basin under water for a couple of hours before being worked over, and I scan the underside of the leaves most carefully for any egg masses. But the immersion drowns most of the flies.

If you are still bothered with whitefly while the plants are outdoors, it is possible to make a teepee with three stakes and a closely tied garment bag fitted over them. In the center of the teepee hang one of those vaporizing strips you can get at filling stations and then group the plants under the plastic and throw earth over the edges to make the "tent" tight. If it is left overnight, the flies will be dead. But do be extremely careful how you handle the vaporizing strip and above all how you dispose of it—it is very potent material.

If plants get badly infected with whitefly indoors and I don't have any of the harmless rotenone or pyrethrum insecticides around, I sometimes mix up a firewater that I learned from one of my gardening aunts. She did it the hard way with a mortar and pestle and a hair sieve; you and I can do it in the blender. In half a blenderful of cold water, pulverize several tablespoons of red pepper, a couple of cloves of garlic, and a cut-up onion. Watch out for your eyes and strain the blended pulp through cheesecloth. The fiery liquid, diluted 50–50 with cold water and sprayed on plants, kills off the hardiest pest. If you add a pinch of dishwashing detergent to the brew, it will stick better to the leaves and stems and be even more effective.

With all this advice, regal geraniums should offer no difficulties for you, and I had better get going and take some of my own advice if I am not to be ashamed of my plants again next spring!

Where Have All the Flowers Gone...

When we bought our house in the country, we found that along with a rather heterogeneous collection of furniture we had inherited a bookcase of garden books that had belonged to the previous owner. Our predecessor at the house had been a distinguished horticulturist, and her books covered a wide range that dealt with every aspect of gardening, and involvement with the land, as it was practiced about seventy years ago.

Since they came into my hands, I have read and reread these books many times, partly to see how gardening knowledge and practices have changed, but also for sheer pleasure. Almost everyone who grows things enjoys gardening books no matter how old-fashioned, for over and above the specific information they contain, they provide glimpses into a way of life that is not only gone forever but already in danger of being forgotten. Sentences that start, "My superintendent informed me that . . ." or "I instructed the men always to . . ." belong to a world that has very little to do with ours. For the sociologically minded they provide an unparalleled chance, not provided so clearly in any other domestic context, to read of the relationship between the talented amateur instructor and the skilled professionals who did the actual labor. Like Mrs. Beeton's book of *Household Management,* old gardening books should be read for their human interest values as well as for any useful instructions they may contain.

But there is one book in this inherited collection that I do not read in such a scholarly or impersonal manner, for I still find it completely practical. It is called *How to Know the Wild Flowers* and was published in 1893. The author was Mrs. William Starr Dana. In these days of increased sensitivity to female identity, I wonder who this highly talented lady actually was—other than the wife of Mr. William Starr Dana—but there is nothing in my old edition that gives any clue in this matter.

In this excellent handbook are recorded, together with many fine line drawings, the wild flowers of the eastern seaboard of the United States, and I have yet to find a more useful handbook.

The book clearly was an immediate success: my copy is an enlarged revised edition published in 1897, and final proof of its enduring excellence is the fact that there is now a modern paperback edition.

It is not hard to see why it was and is still so popular and why I cherish it. Unlike innumerable other excellent books on wild flowers, this was written with the uninformed reader in mind and arranged so that even without any previous knowledge of wild flowers it is simple to use. The plants are grouped according to the colors of the flowers, and the average flowering time of each specimen is included. This means that someone like myself who was not familiar with American wild flowers could see a small pink something in bloom in June and identify it by looking up the color and the time of year, and then checking the possibilities against the description and the drawings. This admirable book also spans the tricky gulf between the ignorant and the highly sophisticated reader by providing both groups with sufficient satisfying information. For the novice mere identification of an unknown plant is often all that is wanted; for the better-informed there are included long plant lists with the Latin as well as the common name and accurate botanical descriptions.

Almost everything I know about our local wild flowers I have learned from Mrs. Dana's book, and I still carry it into the woods with me when I am in doubt. But the process of educating myself about the native flora (as well as the intrusive foreign plants) of this area has brought with it some severe frustrations.

The original owner of my book used to take her children on plant-hunting expeditions, and she and they made it a practice to write in the margin of Mrs. Dana's book the date and place where they found some of the rarer specimens. The record begins in 1898, and one of the attractions of the entries is that they are not all in the same hand. A large family grew up in what is now our house, and apparently there was fierce competition to be the first to spot a rare plant and it was a privilege to be allowed to make the penciled entry. One wavering sentence glows with triumph: the handwriting is unsteady and extremely youthful, but the words leap out, "Rochester Road, 1912 MAYFLOWERS!!!"

I know from correspondence with the youngest of these same children, now herself a great-grandmother, that it was a family custom to pack a picnic lunch and set off on an all-day plant-hunting excursion in a horse and buggy. They used to start very early in the morning and come back in the cool of the evening twilight with the children curled up asleep in the bottom of the buggy. I could reach Rochester in ten minutes by car today, but I shouldn't find those Mayflowers: the place where they used to grow is now a development. But although I have to search rather farther afield, I have been able to keep the book fairly well up to date and record almost all the same flowers those excursionists found with one exception—and the exception is unfortunately a sad one. I have been able to track down almost none of the wild orchids that apparently once grew in this area in profusion. Where the book records seven or eight unusual specimens from "Mary Allen's bog," I have found only one orchid in the swampy land adjoining the now vanished bog, and that stand is decreasing. The book itself records fewer and fewer sightings of these orchids during the later years in which the record was kept, and their decline appears to coincide with the "discovery" of this neighborhood by summer people and the consequent filling in of swamps and marshland for building.

It's no good lamenting what has gone; what lies before us now is to prevent wanton destruction, through carelessness or greed, of what remains. Anyone who knows where rare wild flowers grow would be wise these days to keep this information to himself. This sounds selfish, but very luckily for us there are reservations where we can go and see the missing flowers reestablished and preserved. A natural stand of a rarity is such a fragile matter that the less general knowledge there is of it the better. For, unfortunately, the immediate uninformed reaction to any such discovery is either to pick the flowers or, in more sophisticated circumstances, to return later to try to gather ripe seed. And either action deprives the plants of one of their most important methods of self-perpetuation. There are also those who feel rare plants should be dug up and taken to the so-called safety of a garden. This again is the road to despoliation. Wild plants do not transplant easily; they need special soil bacteria which garden soil usually cannot provide. Digging up orchids, for example, and setting them out in your own yard is just as certain a route to death as if the plant had been thrown away.

The only occasion in which transplanting wild flowers can ever be justified is when the area in which they grow is threatened with a development or road building, and then the work should be left to professionals. Almost every state has a society devoted to the preservation of wild flowers, and it is to them that you should apply if you happen to stumble upon something unusual and fear that it will soon be destroyed.

And there is another action we can take to help the preservation of our lovelier and rarer flowers, and that is to visit and support the reservations where they are protected. We should take our children to let them learn by seeing, just as the children from this house did so long ago. We should give them a chance to appreciate a threatened heritage before it is too late. Family excursions of this sort can be made every bit as exciting as a visit to the beach and almost more worthwhile. For the memory of flowers that will stay with interested children can remain a very happy thing over a lifetime. I know this from my own experience and from a letter from the onetime child who wrote about Mayflowers in my book:

"This must have been when I was quite little and I well remember the joy of uncovering a mayflower and seeing a wild lupin and lovely tasting checkerberries . . . I have so many happy memories of getting wild flowers with mother and daddy."

Petunias and Ivy

Now, when every roadside stand is full of annuals that are on sale, is a good time to pick up a few extra flats of white petunias. Planted by themselves in pots, and kept in a sunny place, petunias will make a pleasant show, but in spite of careful attention, they often get thin and scraggly. However, if a few are combined in a plastic pot with a small ivy the effect can be attractive and quite unusual.

The best way to grow this combination is to crock a six-inch bulb pan and spread a thick layer of rich soil over the internal crocking. Next, center a small ivy knocked out of a three-inch pot, by scooping a little nest in the new layer of soil and setting the root ball into that. If the petunias are in the older or less expensive type of container in which they are not planted individually, cut them apart with a sharp knife. If each plant is left with a neat square block of roots, it will recover quickly from surgery. Nowadays many annual plants are potted individually into little plastic hollows in their container. With these, pop them out individually and there will be no root damage at all. I plant three petunias to each six-inch pot, spacing them in a rough triangle around the outer edge of the pot with a small ivy in the center. If there is going to be a space problem with the root blocks, take out the ivy and carefully work the soil loose from the root ball with the point of a pencil. Without the soil, the ivy roots can be compressed more easily inside the pot and the petunia root blocks forced in without damaging them. The ivy will not particularly relish this treatment and may sulk a few weeks, but if yet more damage is done to the fragile roots of the petunias, those plants may die. With everything in place, fill in all the interstices of the pot with more fresh soil, and thump the pot up and down vigorously so that soil gets in among the ivy roots and no air pockets remain. Now harden your heart and cut off the top two inches of the petunia stems—more if they are very scrawny. This will force them into fresh, branching growth, which is what you want. Water the newly planted pot with a weak solution of water-soluble fertilizer, a little less than is advocated on the instruction sheet, and put a pinch of the same fertilizer into the mister and spray the foliage. Then put the pot in a bright but sunless place for at least a week and don't rewater unless the surface soil looks dry. At this stage some of the remaining petunia leaves will turn yellow and should be picked off. There may also be some yellowing among the ivy, but this is the inevitable aftermath of the shaking-up everything has undergone and will soon cease.

After a week bring the combination pot to a sunny place.

Petunias will not flower unless they are in full sun, and ivy dislikes hot unending sun, so the success of this combination of plants may not seem very probable. But in fact, the petunia foliage grows extremely fast and shades the ivy sufficiently to allow the plant to stand in a sunny place, and until this foliage develops, the ivy can take bright sun as long as it is kept well misted.

The dead petunia flowers must be kept regularly cut off. This is a sticky, fiddling job, but it is the only way to be certain of constant flowers. If possible, display these plants where the rain will not fall directly on them. Rain spoils white petunias; it turns them into a blotched mess. Your plants will look better if their moisture comes from your watering can rather than from the heavens. Once the flowers are in full bloom, feed the pots with full-strength water-soluble fertilizer every other week and try to force additional shoots from the petunias with light tip pinching.

In the fall, when the petunias cease to bloom, cut them out at the soil surface level. This is far better than disturbing the interior of the pot by trying to dig them out. Left without top growth, the petunia roots disintegrate into the pot soil, and the ivy will take over as a winter foliage plant.

I am not overenthusiastic about ivy indoors in the winter. For me it gets ratty and full of bugs, but for those of you who do not want to make a petunia-ivy combination and do like pots of ivy indoors in the winter, this is also a good time to make up a few fresh pots for yourself. This can be easily done by a little cannibalizing of your own ivy and that of your friends. With a sharp knife, cut strands from all the various varieties you can spot outdoors or in pots. Well-grown ivy can always spare a few sprigs without disfigurement. These sprays should be about six inches long and free of pests. Look carefully along the stems for tiny tortoiselike mounds; these are scale insects that suck the life out of plants and often go unnoticed by a novice. Far the best cure is to scrape them off with a fingernail. If there are plant lice on the tips of the stems, wash them off under running cold water. Then sink the sprays under water in a basin for an hour or more to drown those you missed. Next strip the leaves off the bottom two inches of each stem and force these bare ends into a chunk of wet Oasis, the material flower arrangers use which can be bought wherever cut flowers are sold. Oasis can be easily cut to conform to the shape of some attractive vase you may possess, but do this cutting before you insert the ivy stems. Also remember that it takes an overnight soaking to get Oasis really saturated. Unless this is done it will float on the surface of the vase and be unmanageable. Try to arrange the stems of the ivy attractively with all the leaves facing upward and then put the vase with the Oasis in it in a bright sunless place.

At first the effect may not be at all what you expected: some of the ivy stems invariably die and have to be pulled out. But in time all the survivors will settle in, and if the water is kept fresh and regularly topped up, the stems will root into the Oasis. The rooting can be a slow process—I have known it to take months—but as long as you are enjoying the evergreen arrangement of many-textured ivy what does that matter? Roots will have formed when the stems resist a slight tug.

As soon as this happens to all the stems—and different types of ivy may not root simultaneously—lift out the chunk of Oasis and plant it intact in the center of a plastic pot filled with rich soil. To cover the top of the Oasis, it may be necessary to strip a few more lower leaves off the bottom of the new rooted stems, and there should be at least an inch of new soil around the entire outer edge of the Oasis. Don't ram this soil in; ivy likes loose potting. At this stage, I usually pinch out the growing tip of each stem. Transferring the roots in the waterlogged Oasis into soil will call for an adjustment for the top growth, and this is no time for them to be struggling also to elongate. When the roots have adjusted themselves to the new growing medium, the pinched-out stems will branch, thickening up the foliage.

This is another simple way to water-root ivy and transfer it to soil without giving the roots a setback, and some of my most attractive pots of ivy were produced this way. What's more, the variation in leaf size and texture, which comes through using many different varieties for the initial rooting process, produces far more interesting and decorative pots than those available in the stores.

Winter Geraniums

At this time of year, when everything looks so fresh and new, it is hard to plan ahead horticulturally; early summer is not the moment when any of us want to recall the horrors of winter! Nevertheless, this is a very important time to make plans and get going about certain winter-flowering plants, and it is easier to start before the real summer heat sets in and all gardening enthusiasm wanes.

You can, for example, have geraniums flowering fairly well indoors during the winter as long as you have a very sunny window where they can grow and if special plants are properly prepared. This is not the same thing as carrying over the potted geraniums that are just starting to make a fine show on your porch or in your windowboxes. For though these may still look green and appear to be full of bud when cold evenings threaten the end of the outdoor growing season, once summer-flowering plants come indoors the leaves will yellow and die and the buds will fail to open. This is not the result of poor culture on your part but because plants that are used to setting flowers outdoors cannot adapt themselves suddenly to indoor light, which is much less intense no matter how bright the window, nor can they stand the sudden change from the moist atmosphere of the oncoming fall to the dry air of the indoors. But you don't have to throw your old geraniums away; they will bloom for you another season if you will give them the almost waterless rest in a cool place that has already been described.

For winter bloom different plants are needed: they cannot have already exhausted themselves flowering, and they cannot have become accustomed to setting and opening their buds in the open. To get these special plants, buy a few now. The small varieties offered for sale with a single pink or red flower are the best; white geraniums are exceptionally hard to flower really successfully even under the most perfect circumstances. If you can find them, buy the slightly more costly named varieties, for these are the newer hybrids which are tougher and more floriferous than the old, unnamed varieties that still turn up on roadside stands and in Memorial Day baskets.

Once you have bought the plants, plan to repot them; they will need a pot one size larger than their present container. For indoor living, plastic pots suit geraniums very well just as long as there is at least an inch of broken clay shards or roofing pebbles in the bottom of the pot. Florist's pebbles, the white type, should be avoided; they are made from limestone chips which do not suit all plants. If you have no access to broken crockery or roofing pebbles, an inch of coarse perlite can be spread over the bottom of the pot. The danger here is that perlite can spill out through the vents if it gets dry, so it is a wise precaution to cover the bottom of the pot with thin nylon (most households have ruined panty hose from which this can be fashioned) and let that act as a mesh to hold in the perlite yet allow surplus water to drain out. Whatever you use, there will be no problem if you understand that the reason for this action is not a piece of ancient horticultural folklore, but a sensible precaution to prevent earth from blocking the drainage vents. Clogged drainage holes set up a sodden soil condition inside the pot which is disastrous to most plants and particularly to geraniums.

For the actual repotting extra soil is needed. It can be made from a package of commercial soil mix improved by adding half as much damp peat moss and a good peppering of that same perlite. Stir the mixture up like a plum pudding; a plastic bowl is a wonderful help. The texture of the new mix is right when the soil feels fluffy and does not lump into a sodden mass in your hand when you squeeze it. Put a pad of the mixture over the drainage material and then knock out the young geranium, not by pulling it by the neck but by turning the pot over, spreading your left hand across the top and rapping the edge of the container sharply against something hard. If the plant does not slide straight out, thump the side of the pot again; sooner or later the entire root ball will disengage itself from the internal sides of the pot and either drop into your waiting left hand or allow you to pluck off its original pot with your right hand.

Set the little plant in the middle of the new pot and dribble the soil mixture between the side walls of the pot and the root ball. This extra soil should be rammed down very firmly; something flat and narrow is best as a tool; I use a paint stirrer if I have one handy, but the handle of a fork will do in a pinch. This hard potting is essential: geraniums never flower well if the soil is loose around their roots.

The next stage is to recycle the repotted plants into winter-flowering specimens. This is done by putting them where they will get at least half a day of full sun and regular watering. It is a constantly repeated fallacy that geraniums need to be kept dry. This is not so and will be disastrous to the appearance of the plants you want to look well in the winter. A geranium kept short of water loses its lower leaves, and you will end up with unattractive feather dusters. Cut off the existing bloom and nip out every bud. Continue this treatment relentlessly until the end of August, and add to the punishment by pinching out the tip end of each shoot as soon as it elongates to about four inches.

This frustrating treatment, for the plant will fight continuously to be allowed to produce buds, turns all its energy into producing strong, compact foliage. As this thickens up, give the plant an occasional light feeding: a water-soluble fertilizer at half the recommended strength every second week is quite enough.

Don't worry if you can see roots at the drainage holes; geraniums flower better when potbound, and this should not be the signal for you to pot them up again. But if you want a huge indoor plant, you can repot two pinched-out plants into an azalea pot, using all the same precautions about drainage and rammed soil. This will produce some yellowing of leaves in the center of the combined plants which should be picked out to allow light and air to get in. As long as the pot with combined plants is twisted around regularly so that all the foliage gets sun, it will turn into a dramatic winter-flowering plant.

After mid-August stop all the pinching and allow the buds to develop. As these straighten and plump up, if the pots have been outdoors, take them indoors to a sunny, open window. This will give them a chance to acclimatize to indoor life before the heat comes on and it will also enable them to open their first flowers inside. The frustration that winter geraniums have undergone all summer makes the first flush of bloom relatively certain, and if this stage can take place on a windowsill, you will have a far better chance of continuing flowers after the strong outdoor light begins to weaken.

No matter how good a grower you may be, winter geraniums will never be as striking indoors as outside. But you can take some comfort from the fact that they don't do very well in greenhouses in midwinter either—not, that is, at the heat most of us run our greenhouses. Geraniums are heat-loving plants, so they are not very satisfactory deep-winter greenhouse plants.

This is one of those occasions when the indoor gardener with a warm room and a bright sunny window may do better than the owner of an elaborate setup under glass. And for most of us, there's something pretty satisfactory about that too!

I Remember, I Remember...

For many people the scent of certain plants can revive memories with a vividness that nothing else can equal, for the sense of smell can be extraordinarily evocative, bringing back pictures as sharp as photographs of scenes that had left the conscious mind. Whenever I come in contact with flowering ivy, I find I have almost total recall of a very early ground plan of my parents' first country garden. That particular garden was later entirely redesigned as my parents developed new tastes, and when I think back consciously about how and where they managed their plants, I always visualize the details of the later stage, which I knew far longer and also when I was getting interested in plants. But the instant some dormant lobe in my mind is retriggered by the pungent, heavy, almost unpleasant smell of flowering ivy, I suddenly see with great clarity the first garden including minute details—with which I sometimes confound my brother, who does not possess this visual sense of total recall. Above all, I remember the source of the ivy memory jogger: a small peat shed, my special hiding place during the early years of our occupation of that house, which had to be taken down around about the time I was nine years old. It was so weighted down with luxuriant ivy that it was judged unsafe.

In New England, except for the Baltic variety, ivy rarely flowers; the harsh winters kill it back too hard. For ivy must scramble to a certain height above the ground and reach a fixed stage of leaf maturity before it will set a bud. And it was not until I recently brushed against a flowering spray, in a warmer section of the country where ivy flourishes all too happily, that I again realized that the old nerve still existed and was only waiting to be plucked. Tree ivies, by which I do not mean the modern hybrid of ivy and a fatsia, called fatshedera, but true ivy grown into a standard form, are no longer popular. They used to be very fashionable and were used as what then were called "dot," or to us "accent," plants to give height to an area planted entirely in low annuals. True ivy standards are now hard to come by partly because they take time and patience and a good deal of room for winter storage, but also because they can be made only from cuttings of ivy that are mature enough to flower, and rooted cuttings of old wood are always hard to strike. I have no passionate wish to possess a tree or standard ivy; I suspect I would have just the same trouble with it as I have with the juvenile trailing form we all grow. But I had almost forgotten their very existence until that pungent smell also reawakened the visual memory of my grandfather's long parterre beds with the ivies standing like sentinels at regular intervals along it.

Unlike many people, I am not all that enthusiastic about roses. I love climbing roses and grow a lot of them as well as old-fashioned rambler roses, but in a very simple style scrambling among the trees or flinging themselves over stone walls where they give no trouble except to require an occasional thorny cutting out. But I am not much interested in rose gardens, and I dislike the heavy scent of roses indoors. I can rationalize this by saying that roses in this country take too much care and are prone to too many diseases, while roses indoors shatter irritatingly fast, all of which is perfectly true. But I think the truth of my indifference has deeper roots. The family rose garden, which was my father's delight, was always at its superlative best in June, the time of year when vital examinations were held in English schools. I was rather a ten o'clock scholar and used to leave a great deal undone until the last possible moment. Then I used to pace around and around the circular rose garden, which was a little set off to one side and therefore a quiet place, feverishly cramming for the upcoming examinations. The scent and sight of roses have therefore an association of tension in the back of my mind, and this is probably the real reason for my lack of interest in them.

And I find this childhood association syndrome exists in all sorts of gardening matters. Our country garden is surrounded by tall hedges in which we battle the twin plagues of bittersweet and Japanese honeysuckle. In spite of our counterattacks both plants are going to outlast us, and where the bittersweet is concerned I find I have ambivalent feelings about cutting out the fruiting strands even though they are doing their best to strangle us. As a child I used to hunt for bittersweet berries in the hedges along the chalk downs. Those two-tone sprays that I now cut out were there a trophy to be proudly carried home, for bittersweet is not the nuisance in England that it has become in some areas here. I am sure it is a relic of this feeling that leads me to emphasize the importance of the berries for the bird population and turn my fiercest efforts against the unberried strands.

I deplore the pythonlike grip with which the Japanese honeysuckle strangles everything within reach, but I love the elusive fragrance with which it envelopes the country garden at this season. To me that epitomizes everything we enjoy about this slightly ramshackle but deeply loved garden, and that same scent, it seems, is apparently going to affect my children just as ivy keeps its hold over me. For recently standing on the porch in the evening with honeysuckle perfuming the air, a visiting daughter who now lives where this vine does not grow suddenly remarked that to her the smell of honeysuckle would always mean that school was out and the endless sunny summer had begun. She had forgotten the sensation until that very moment.

Scents bring memories, and many memories bring nostalgic pleasure. We would be wise to plan for this when we plant a garden.

The Intruders

The wild flowers are looking particularly lovely now along the verges of the back roads and even on the banks of the new motor highways. As an ex-alien myself, I am always amused at the number of non-American plants that have managed to get themselves accepted as local inhabitants and the arrogant manner in which they flaunt their takeover abilities. Most of the flowers that catch the eye as we drive around the New England countryside at this particular season we take so much for granted that it is hard to believe that they have not been here since the beginning of time. But true American wild flowers that shared this countryside with the Indians long before anyone's ancestors scrambled onto the *Mayflower* or set sail for Virginia appear in force later in the year, their time of glory is high summer and the fall. The wild flowers that we enjoy in early summer, the lovely spreading sheets of oxeye daisies and the pervasive yellow hawkweed, for example, came in with the white man from Europe and are not indigenous to this country.

Day lilies, which are such an endearing sight in abandoned gardens and neglected cemeteries and the ditches beside the roads, are also invaders—this time from the Orient. The sweetly scented lemon day lily (*Hemerocallis flava*) is still quite rare outside old gardens, but the same cannot be said of its far more aggressive relative the tawny day lily (*H. fulva*), which has spread everywhere that suits it and in the process has shouldered out everything else in the neighborhood.

In their native land day lilies are considered particularly delicious gourmet magic, and almost every part of the plant—the flowers, the peeled stems, and the freshly sprouted shoots—is eaten. Recipes for cooking flowers are now turning up in many modern cookbooks, and since the tawny day lily is too coarse and invasive a plant to be really welcome in a small yard, using it for food may be a way of eliminating it from your garden. For it has been my slightly unkind observation over many years of horticulture that if a useful economic purpose can be found for a plant it nearly always then becomes something of a rarity! So if you are troubled by big patches of *H. fulva* and want to grow the newer and more delicate hybrids that the plant breeders have now provided for us, have your family cultivate an insatiable appetite for every part of the tawny day lily; this, I can almost guarantee, will turn it into a rarity in your yard!

Queen Anne's lace or cow parsley, which is in fact a wild carrot, also came in with the white settlers, though not by accident; it and, to our eyes, the more useful edible variety were deliberately imported. The European herbalists of the sixteenth and seventeenth centuries were greatly involved with the use of wild carrot for medicinal and erotic purposes, and what is now a delightful, though rather obtrusive, weed had an active place in pharmaceutical practice. Still other Europeans are the ragged robin of the woods, the yellow field tansy, and the number of common weed plants in America that were either deliberate or fortuitous introductions, are by now too many to be listed.

On the whole we have been the gainers by this wave of immigrants—although I would not for a moment suggest that we have been anything but wretchedly treated by some of the diseases that they also brought with them. Nevertheless, the countryside now would look extraordinarily bleak without these foreigners. But some are an increasing problem, for the natural factors that keep them in balance with the other local flora in their native land are not active here. One example of a gravely increasing menace is the Japanese honeysuckle. This is turning into a killer of small trees along the eastern seaboard, and the outlook for reversing the trend is not good. Japanese honeysuckle twists its way up a tree in such close contact with it that the use of systemic weed killers introduced into the plant system through the roots or herbicides (foliage killers) is not practical, for these materials would inevitably kill the unwilling host as well. To clear an area of this pest, the plant must be torn out of the trees by hand, and the running roots dug up—and where are people to be found in sufficient quantity to do such work today?

We have also been short-changed about weed pests in our lawns. Almost all the horticultural nuisances that make grass growing such a problem in this country are foreigners. The pestilential ground ivy is an Asiatic, while couch grass, crab grass, shepherd's-purse, dandelions, docks, and plantains are all European imports. Just think of the unblemished velvety smoothness of our lawns if we could but deport all these particular weeds!

But the traffic has not been entirely one way; there are plenty of American weeds rampaging around in other continents. In England the Virginia creeper is so much at home that were it to speak it would have a British accent. As for its look-alike dangerous friend poison ivy, persistent efforts, believe it or not, were made to introduce that into English horticulture from the seventeenth century, when berries were sent back, right down to the end of the nineteenth century, when common sense about this particular plant seems at last to have prevailed. How and why this menace failed to establish successfully in England I do not know, but it certainly was not for want of trying!

It does seem a little unfair that foreign plants turn into pests so easily in this country while our local nuisances either die out or become tamely domesticated when they emigrate. So although plant quarantines seem intolerable, experience has proved that they are immensely important. The pity is that they were not established even more strictly much earlier.

July

Biennials

June is the time normally suggested for sowing the seeds of biennial flowering plants in this climate. Biennials are plants that make only leaf growth the first year, hold a sturdy green rosette of leaves all through the winter, and throw up flowering stalks in early summer the following season—flowers that appear at an extremely useful moment when the first spring flush of blossom is gone and the hot-weather annuals have not yet taken over.

Biennials call for a lot of work, and their winter survival is not absolutely certain in New England. In fact, I once got so discouraged about them that I ceased to grow any. But by depriving myself of sweet williams, Canterbury bells, pinks, and foxgloves I cut myself off from one of the big flowering periods in the yard, and I now have taken to them again, but on a rather more relaxed pattern that is slightly less demanding and still seems to produce the same odds about wintering over successfully.

One of my changes of pattern was to sow the seed later than is customarily advised. This, I agree, means that germination has to take place during the heat of the dog days, which always makes difficulties. But as a result of slightly later seeding, the plants do not have long enough growing weather to turn into immense cabbagelike objects, and it was my experience that enormous plants are the least likely to make it through a bad winter. I now sow seed from early to mid-July, and I use inexpensive packages that I pick up at our local garden center. I used to send away for special seeds from distant or even foreign nurseries, but with such chancy plants I no longer consider this expense worthwhile. I grow sweet williams in a single color called Newport pink, a watermelon shade that blends in well with everything. Avoid a package of mixed colors in sweet williams, for these will give a spotty effect. If you do send away for special seeds, consider getting a package of auricula-eyed sweet williams (*Dianthus barbatus auriculaeflorus*), which are tall and highly spectacular and also do well indoors as cut flowers. With Canterbury bells, I always buy the cup and saucer type (*Campanula medium calycanthema*) and hope that there will not be too many blues in the packet; the pink and white varieties are what I really prefer. But if a blue does burst upon the scene, never forget that this color forms a strong accent point in any grouping. Foxgloves can be found in the new Excelsior hybrid form which flowers all around the stalk and is therefore a better buy. Incidentally, all these plants are European imports, but only the foxglove is native to England. Sweet williams and Canterbury bells, despite their delightful country-style names, were introduced into England from Europe in the sixteenth century.

The seed can be sown in the open ground, which I find the most hazardous choice, or in a raised cold frame or in seed pans. Make sure that the soil is flat and firm, and do not cover the seed too deeply. Shade the bed from hot midday sun with an old sack raised slightly on sticks to allow circulation of air or wet newspaper raised on large stones. Hot-weather germination is usually poor, because the ground gets baked out just as the seedlings are breaking their outer casing. Shading until the cotyledons, or first seed leaves, are above ground prevents this difficulty. It is essential to keep the ground moist, and this naturally attracts slugs. For outdoor seeding, a shallow saucer of stale beer to attract slugs is a necessity. As part of easing my struggles, I have taken to sowing the seed in small flats which are placed on a shady shelf in the cool shed; I seem to have quicker, thicker germination this way. As soon as the seeds are above ground, they should be slowly accustomed to full sunlight and ruthlessly thinned. There are very few gardens that need hundreds of these little plants; two dozen of each will be more than most of us can handle. With Canterbury bells try to keep the small delicate-looking seedlings, if you prefer the white and pinks. The blue varieties show as much bigger plants from the very start. After thinning allow the little plants to develop with careful attention to regular watering and an occasional dose of liquid fertilizer to speed them up.

Once the biennial plants have developed a rosette of leaves that is about an inch and a half across, they should be moved into rows in a sunny open place—our plants are spaced a foot apart in the empty areas where early vegetables have been pulled out. Firm each plant in carefully and water it individually, and then mulch the area under the leaves and along the rows very thickly to conserve moisture and keep the disturbed roots cool. Usually they will move without turning a hair, but if the weather is extremely hot, a light misting from the hose early in the morning, so that the foliage has plenty of time to dry off before nightfall, will revive wilted plants.

Once in these rows, biennials should take no more of your time while they slowly expand. Don't, however, try to push them into super growth with fertilizer. You want sturdy, tough, rather small plants with strong roots as the cool weather sets in, not lush, floppy specimens.

After the first light frost has killed the summer annuals—or, if you clean up your flower beds earlier, any time after Labor Day—the biennials can be moved again, this time into their final flowering position. Don't take every single one from the growing area; leave some as insurance. This is a very critical stage in the survival of the moved plants. If there is a long open fall, and the plants were not too sappy, they will move without faltering and put down fresh roots in the still-warm soil that will carry them through a normal winter. But if fierce cold weather sets in early, you may well lose all the plants you shifted. That is why some should be left growing in the original rows.

After the ground has frozen, put a pad of dry leaves under each green rosette. Biennials often die because their leaves get frozen onto the ground underneath heavy snow, and a mulch under the leaves allows for air circulation. Don't spread winter mulch on top of them. Since their leaves remain green, some activity takes place all the time, so the leaves must have air and light.

In an average winter biennial plants stay in excellent condition until around late February, and then, just as you are congratulating yourself on their well-being, disaster strikes. As the weather warms up, the soil surface begins to thaw slightly by day and then refreeze at night and in so doing tears and heaves the roots of plants. Biennials root just under the soil surface, and it is the early spring heaving that so often does them in. If you have spare unfrozen mulch, now is the moment to spread it over the ground around the biennials and put even more under the green leaves. This will keep the ground solidly frozen and save the root system.

Once the frost finally leaves the ground, pull the mulch out from under the plants and pick off any leaves that have died. Almost every plant will have a few yellow leaves—if they are around the outer edges all is well—but dead leaves in the heart of the plant are a sinister sign. As the weather warms up, the plants will stool out, which means expand sideways, and before this happens dead plants should be replaced with survivors left in the original rows. Moving these once they have begun to expand gives them a severe shock from which they do not recover easily (one of the reasons why biennials are not a good spring buy at a garden center). Plants that have pulled through the winter turn into very large leafy specimens that offer considerable wind resistance. By May, when the flower stalks appear, stake the strongest of these to hold the entire plant steady.

The show of flowers when it comes will be spectacular, and if you faithfully remove dead flowers from pinks and Canterbury bells you will have three flushes of bloom that will last into mid-July. Once the flowers cease to come, pull out the plants, for their lifework is done.

Foxgloves, which grow by nature in woods, do quite well in a sunny place, but if you have a choice in your flowering area try to give them part shade. These plants are great self-seeders in a curious on- and off-again manner. Some years after you have grown foxgloves there will be no seedlings; then suddenly you will be overwhelmed. Sometimes they appear several years after you have ceased growing the plants at all! Small foxglove seedlings are always worth using, for they produce excellent flowers, but any self-sown seedlings of other biennials should be discarded, for the colors will be poor and muddy-looking.

In climates milder than New England I can see no excuse for not growing these excellent plants, though I still would not sow the seed as early as advised in most books—which incidentally is English practice for a very different climate. Here, in spite of some disappointing years, I find biennials still worth the gamble, for if and when they come through there is nothing to touch their long-lasting loveliness.

The Salt Marsh

We own a small piece of salt-water marsh in which is epitomized all that is delightful and vital about our shore wetlands and the importance of preserving them. And when, with the very best of intentions, I nearly ruined one of the great assets of this piece of property, I was also taught by our little scrap of coastline the extreme fragility of the ecosystems and food chains that are supported by areas of this type, how easy it is to disrupt them, and how long it takes for any such disturbance to be repaired.

The trouble began some years back before there were any laws on our local or state books prohibiting changes in wetlands. I decided it would be pleasant to carve out a small sandy beach for the grandchildren along one small corner of the salt marsh, and to do this there had to be a little dredging. This was carried out with great care under my supervision, and I took tremendous pains to see that the operation did not disturb or infringe upon any other part of the marsh.

All this effort was not only absurdly expensive but fruitless and disastrous. The work was done in very early spring, and the first strong moon tide that surged in after the sand had been laid down washed it all out to sea, leaving a jagged hole filled with seaweed and other unpleasant ocean-floor debris in its place. That was the end of the beach, for the operation of getting the sand into place had been far too complicated to be undertaken a second time. And if that were not trouble enough, by midsummer I discovered that, in spite of all my care, I had in fact seriously disturbed the delicate balance and interaction of plant and animal life that must coexist if a salt marsh is to survive. For after the sand was washed out over them by the pounding tide, the marsh flowers and all the other vegetation died, to be replaced by foul-smelling, slimy mud which formed a quaking quagmire without the stabilizing roots of the marsh plants to keep it firm. This mud was deep enough to be dangerous for small children, for they sank up to their waists and could not extricate themselves without help. So the immediate aftermath of my attempt to give the family a pleasant place to swim and sun was the need to post the area as being dangerous because of the deep mud.

At the same time, and obviously coincidentally, the huge beds of clams that had always been plentiful along the base of a nearby sea wall vanished; worse still, the mussels, the source of much delicious soup, that had clung thickly to the timbers of an abandoned wharf in the same vicinity moved away, possibly because the eel grass had been killed. Only this summer, ten years later, have we at last spotted a few mussels again tentatively attaching themselves to the old pier.

The mussels have been the last of the local inhabitants to reappear, but the regeneration of the spoiled area was all very slow. It had to take place in a slow, orderly sequence as successive life support systems reestablished themselves and then, in turn, made it possible for the next species in the ecological chain to follow.

There was very little we could do to speed up the process. Our only contribution was an attempt to stabilize the muddy sea floor by removing the evil-smelling seaweed out of the cavity that had been the potential beach and filling the space with stones, which provided a base for a new sea floor that was slowly washed in by successive tides. And as this stage progressed over the summer, the smell of decay that had hung so heavily over the entire area slowly diminished. The reeds were the first plants to return, but they crept in very slowly, taking almost four years before they re-covered the exposed floor. Now when the tides slide slowly in over their waving foliage, shoals of small fish come in too, swimming untiringly in elaborate patterns that never collide and feeding on the algae and minute marine life among the reeds.

The roots of the reeds restabilized the mud, and as that hardened up, we began to find crabs again at low tide and shells preempted by hermit crabs. By this time jets of water, and a little cautious harvesting, proved that the clams had also returned, happily bringing with them oysters that we had not found previously. To a casual eye the reeds and mud flats at low tide may still seem barren and dead. But we who have watched over their slow regeneration know that in fact they are again pulsing with life.

It is frightening to realize how greatly I imperiled not only the rather rare marsh plants that live poised between the land and the sea, but also the chain of marine life that in turn a great deal of sea life supports. If that chain ever is severed on a large scale, we will lose many of the fish that provide us with employment, food, and recreation.

I had to learn this fact the hard way, and I was lucky—but have we as a nation yet learned this lesson?

Water

We are fortunate enough to possess an old well hidden so unobtrusively under a large copper beech that we lived several years in the country house before we discovered its existence. Fortunately, it was not a conventional well with a cover that can rot through age (the kind of well that has caused some fearful accidents when its whereabouts have become forgotten) but pipes that had been bored down until they tapped an underground freshwater stream. The water was raised by means of a tall windmill and used both in the garden and in the house. Since every day was not windy, large storage tanks to hold a surplus supply existed in what is now our toolshed but is still called the "pumphouse."

Last week, when the all too familiar postcard arrived from the local authorities forbidding the use of outdoor garden hoses except on specified days, and even then at impossibly early hours, we reconnected the machinery that now brings up this deep water by means of a small electric pump, and after a lot of priming—a once familiar skill at which I now seem to have lost my touch—we again tapped our personal supply of water.

When the hose water is forbidden us, it is vitally necessary to have extra water for the porch plants, the house plants that are summering outdoors, the vegetable garden, and all the seedlings. Before I knew of the existence of our well, water restrictions did serious damage to all these gardening activities. Potted plants cannot exist long during summer hot weather without regular supplementary water, and seedlings can wither away in a single day. Although the vegetable garden has a thick, moisture-retaining mulch, this gets constantly disturbed as the finished rows are pulled out, and the speed at which exposed soil dries out can stunt the growth of the upcoming crops as well as ruin their taste. Vegetables need a regular heavy supply of water to grow well. This is something that should always be remembered about trying to raise vegetables, now that so many people are going back to the old idea of having a bountiful home plot. Unless there is plenty of readily available water in addition to that provided by heaven, any dream of this sort will be a great disappointment.

Our underground water supply is very cold, for it lies deep below the ground. The stream from which it is taken runs parallel to the shore, and when we first came to this neighborhood there was a line of ornamental well heads marking its route in many of the neighboring gardens. Our ignorance about the existence of this water in our yard arose because the former owner never indulged in any such elaborations. The two main wharfs where whaling ships once discharged their cargoes may have been built where they still stand partly because this steady supply of sweet water was so near, for the ships refilled their water casks from wells that still exist. When eventually the town laid water pipes, many of the old wells were sealed up, even in my time, two have vanished, for until fairly recently none of us ever dreamed that the period was so close at hand when the available piped water would not be sufficient for every use.

But in recent years, as the town has grown, the search for more water for the general supply has become extremely urgent, and the restrictions on casual use or overuse have become sterner and sterner. In consequence a great many of the old wells have been reactivated, and gardeners without one regard them with great envy. As I see more and more of the original wells coming back into use, and even new ones are being dug to tap the same source, it has worried me a little that the old supply may prove inadequate to the demand, but so far this has not happened.

The cool freshness of the well water is very unlike the taste of the heavily chlorinated water that comes out of our faucets for it is quite rare to taste water without additives these days. But these virtues which I find so pleasant do not seem nearly so attractive to the plants; they resent the chill the well water gives to their roots and immediately show it by a certain amount of yellowing of the leaves. I also do not enjoy dragging enormous lengths of hose across the yard, which must be done to get the supply to the vegetable garden, and the innumerable pinhole leaks that exist in the ancient coils of hose that drench me with icy water as I walk by. But both the plants and I have to put up with such rather minor disadvantages, for without the well a great many cherished treasures would die.

The one group of plants that simply cannot take a cold douche from the well are the tiny seedlings. Potted plants may show resentment, but they survive, and the vegetables continue to develop unmoved. But seedlings that have just germinated are killed by such cold water. For their use, therefore, I fill a large plastic garbage barrel with well water and allow it to warm up for at least a day in full sun before dipping it out for use with a watering can. Hand watering is an added chore at a busy time, but it brings the side advantage of forcing me to look closely each day at every pot. In that way unwanted weeds are immediately taken out before they choke the special plants. Watering seed pans by hand, particularly if I remember to put the fine rose into the nozzle of the watering can spout, also prevents the soil from packing down—which often happens with a carelessly brandished hose. Proper watering is often overlooked in handling seedlings; indeed, many watering cans are no longer sold with a rose, an extra detachable nozzle with the business end perforated by a great number of minute holes, so that the water comes out in fine threads. This prevents seeds that are not yet up through the surface soil from being washed around and also is a much safer method of watering small seedlings themselves. If your watering can has no such rose, seed pans are better handled by being set into a basin of water and left there until the soil surface becomes moist.

Slightly warmed water is better for all potted plants, not only seedlings, at all times of year. Those fortunate enough to own, or be thinking of owning, a greenhouse should have hot and cold water brought to it with a mixer faucet so that warm water is always available. Lacking this facility, if there is space, it is a good idea to have a big plastic barrel inside the greenhouse which is kept filled from the hose. This will warm up to the same temperature as the greenhouse itself, and water from it will help plants better than an icy flow from a hose. A barrel filled with water inside the greenhouse also adds invaluable humidity. The warmth of the water is not the only reason for being rather careful about using water straight from the hose on either house or greenhouse plants. Today there is so much chlorine in all water to make it safe for drinking that the level can damage some plants. Water that has stood overnight in a wide-mouthed container will be free of chlorine, which will have evaporated.

Rainwater used to be considered the best possible water for plants; for this purpose, many houses had big wooden water butts, open in England but covered in America, into which the gutters drained. Today such rainwater may not be a panacea for potted plants: we read such horrifying accounts of the impurities and toxins carried by precipitation that we cannot mourn the passing of the water butt. But its replacement, a plastic barrel filled with tap or well water that is allowed to stand for twenty-four hours, is an excellent substitute—with one warning: the barrel must be completely emptied and inverted for a couple of hours every few days to dry it out. Only with this treatment can you be certain that you are not providing, in addition to warm water for your tender plants, a delightful subtropical breeding ground for the larvae of mosquitoes!

Naturally

There are occasions when all of us feel a wry kinship with Job and wonder what we have done to be so singled out by misfortune. But I go a little further than that: horticulturally speaking I must, I think, be the lineal descendant of Jonah as well as Job, for disaster and foul weather always seem to haunt any garden occasion with which I may be associated. This has happened not once but many times.

The year my greenhouses were open for inspection, the shelves carrying the forced bulbs broke a week before the event, ruining not only the bulbs but also some special lilies underneath them that I was forcing for a family wedding. The last time my plant windows were on display, the plastic blind in front of one of them fell on me in a suffocating veil just as I was demonstrating its carefree qualities. In its descent, it knocked a great many pots off their staging tubes, and an enthralled group of spectators watched as cyclamens and primulas crashed down onto the rug in a hideous mess of earth and broken leaves while I struggled to free myself of the caul that enveloped me. That day was also drenchingly wet, and there must have been several hundred pairs of rubbers dripping on the parquet floors.

Recently my garden was open, and I labored hard in advance on the display porch and the flower-growing area. For although the place is far from a show area, I did at least want to seem to have tried! The day before the tour, there was unending rain that battered the flower beds, reduced the petunias to slime, and knocked down everything that had not been tied to a stake. But there was worse to come. That night the wind shifted and howled with gale force straight onto the porch where all the container plants were staged. All night my sleep (which was, as you can imagine, restless) was punctuated by the crash of falling pots.

The great day itself was unexpectedly sunny, but the porch was a shambles with almost every pot knocked down. Normally it takes me two or three days to stage that area; this time I got the reserve plants, all of which had to be carried from a considerable distance, into position in a record staging time of four hours so that we were ready for action by the opening at ten o'clock. It was fortunate that I had extra plants, but they were far from my best specimens. The overall effect of the porch was not at all what I had intended—or what it had looked like the day before.

The flower garden had also suffered dreadful damage, but aside from some emergency propping up, nothing could be done, for the first cars were arriving while my daughter was still at work. And the first visitors were already at the far end of the yard before the men who help me with the mowing, who had been hastily summoned, had had time to saw up and pull away an enormous bough of our old elm which had crashed down right across the small vegetable plot.

On the whole, the day was a success; people undoubtedly enjoy looking at the disasters of others. But as a household we did not enjoy it; our pride was as badly battered as the yard.

With such an outstandingly fatal record, maybe organizers of all future events will take warning and stay well away from me. Either that or sell the tickets in advance, for I am not prepared to swear that I will ever have anything worthwhile on display!

Vacation Time

Do you dread the family vacation because you can't imagine what is going to happen to your cherished house plants or carefully tended yard? Nothing can replace the owner's eye for the very best care, but there are some tricks that can help you relax on holiday and not worry about your plants too much—for a limited period.

For house plants there are alternative methods of handling. You can buy a quantity of thin plastic garment bags or medium- or large-size plastic garbage bags; in either case get the kind that have no holes in them. You will also need a package of thin bamboo stakes bought from the same hardware store that sells the plastic bags. Get the stakes quite long; you can always cut them down, but you cannot magically extend them. They need to be longer than your tallest potted plant that is not a huge individual accent plant.

Take your plants to a bright but sunless position: it is vitally important to get the plants out of the sun but still keep them in good light. To use a garment bag, tie one end tightly with an elastic band, then insert two stakes into each far corner, using empty cotton spools like candlesticks to hold the stakes. Water the plants and put them into the garment bag, put two more stakes into the near corners held up the same way, and then tie the end of the bag tightly, thus enclosing all the plants in an elevated plastic tent. The stakes need only be tall enough to prevent the plastic from resting on the foliage of the tallest plant, and you can puff the bag up like an oversize balloon before fastening it if you need more internal space. Plants will survive a long time under these conditions, and you can construct as many airbags as you need. If you have only a few house plants, it may be simpler to put three stakes into each pot, slanting them outward in a triangular shape; again the stakes must be taller than the tips of the foliage. Slip the watered pot, stakes and all, into a plastic bag that is large enough to tie tightly above the top of the stakes, and again stand the shrouded plants in the brightest possible place where sun will not strike the plastic, for if it does the air inside will heat up and the plant will die.

If you own enormous accent plants that no bag can entirely contain, water the pot soil thoroughly and then encase the pot itself in a plastic bag. Don't let the bag be too form-fitting; there should be air space between it and the pot sides. Wind the top of the bag tightly around the stems of the plant, which will stick up through it, and seal the plastic as closely as possible with electrician's tape; the aim is to prevent the loss of moisture from the soil beside the stems. If you keep an air conditioner rumbling in your absence, put the plants in the same room; they will last better in a cool atmosphere.

You must plan a little more ahead for the yard. About a week before you leave, thoroughly weed the flower beds and then let the hose run slowly all night so that the soil is saturated. Next lay a four-inch layer of mulch—that is, some inert material that does not contain weed seed—to cover all the exposed earth. Spread it close to the plant stems, but not touching them, and under the mats of creeping plants. Mulches can be bought at garden centers. I happen to use fir bark, for that looks pleasant and lasts well, keeps the soil moist and cool, and lets rain through. Spread thickly: any mulch also inhibits weeds. There are all kinds of possibilities for mulch—including grass clippings from the lawn and even old newspapers—and the choice is yours. The only material I would not use is peat moss, which is sold bone dry and is extremely hard to get evenly wet. If it is laid down dry, it will steal moisture from the soil while simultaneously preventing any rain from passing through to the parched earth underneath. If somehow you have got it wet, peat moss dries out extremely fast in hot sun, with all the noted disadvantages and the additional problem that it will shrink as it dries and leave exposed patches of earth where weeds will spring up.

The day before you leave, cut off every bloom and bud that shows color from all your plants, even the roses (other than ramblers), and shear back sprawlers like petunias and nasturtiums. The bereaved plants will make a fine new set of buds to greet you on your return.

Mulch the soil of your vegetable patch in the same way, but invite a neighbor to pick any ripe beans, eggplants, squash, or tomatoes. Unless these are kept harvested the plant will cease producing. In return, ask the same neighbor to water any potted plants that are standing outdoors in the shade; there is no other way of keeping these alive while you are gone.

And if you have made the mistake of starting to water the lawn, you will have to ask him to attend to that chore too. A lawn that has been watered regularly cannot quit cold turkey. But do not suggest that the flower beds be watered; the mulch will hold in sufficient moisture for the roots, and overhead watering during the dog days brings on mildew.

No one does all these things, but try to do some of them. They can make a great difference to the homecoming.

Let's Not Spoil It Again

During the past few years there has been a wonderful resurgence around here of the many wild things with which man shares the living world. Nights have been brilliant with the flashes of fireflies, long absent in such quantities; there are unusual birds in every yard—we, for instance are hosting a pair of southern mockingbirds—and wherever you look, the flowers and hedgerows are alive with swarms of butterflies in flashing colors. I am delighted to see all this, but I am particularly happy with the return of the butterflies, clouds of colors which represent the reemergence of a native that it was feared might have been lost forever through the overuse of powerful sprays.

As a child, I can remember finding up in the attic a small wooden cabinet with glass-covered drawers, containing my father's collection of moths and butterflies. Each was laboriously identified in a childish handwriting and carefully displayed to show off the wing markings; but their bodies were impaled with pins, and I can also recall my disgust at this dreadful sight. My mother, hearing me slamming the drawers shut, carefully explained that my father had made this collection many years earlier, when catching butterflies and robbing birds' nests of their eggs was considered a proper occupation for small boys, just as we now encourage Little Leaguers. It was also made quite clear to me that my father always killed each catch in a bottle of chloroform before pinning it down. This took some of the horror off my mind, though I still refused ever to touch a specimen, but I often checked them carefully, for they helped me to identify the many butterflies that flew about the garden.

But although my father was rather proud of both his collections—there was also one of birds' eggs—we, as children, were strictly forbidden to attempt anything of the kind ourselves. Some of our contemporaries still went in for these practices, but my parents by then strongly disapproved.

Nowadays thinking people try to train their children to preserve rather than destroy living things, but we do so in a general rather than in a specific manner. Very few of us would dream of encouraging a child to rob a nest or chloroform a butterfly as it beat its wings against the glass sides of the bottle. Yet we remain curiously uninvolved about the use of twin demons of destruction, aerial and herbicidal sprays in our neighborhoods, although these are far more lethal to birds and insects than the casual depredations of children.

When I first came to this country I was amazed at the variety and size of the butterflies. But after overhead spraying got under way in a vain attempt to control mosquitoes, butterflies became few and far between except for the ubiquitous cabbage white variety, which seemed to be immune to everything. The outcry raised in recent years about the overkill of aerial spraying has led to more thought being given to the matter, and we no longer receive an automatic drenching without notice. And since there has been less spraying, the butterflies and birds have made a comeback. Through better luck than judgment, we have in fact been given another chance to retain one of America's more delightful heritages, for nature will forgive the outrages we visit upon it given any sort of chance. But it is up to us to make certain that any such second chance is not wantonly thrown away again.

The aftereffects of more restrained policy when poisonous sprays are involved are everywhere. It is, for instance, going to be an outstandingly beautiful year on the waterfront, because the sea lavender, which is butterfly-pollinated, has spread into huge shimmering patches that show as a glint of silver against the blue sea. Because the land has been free of airborne poisons for several seasons, the stretch of marshland I nearly ruined, which has had a long journey back, is making great progress in a sudden burst of vitality. Tiny cedars are competing with prickly little hollies just above high-tide mark, and both are struggling against the intrusions of the purple loosestrife. Nearer the water the scene is dignified by huge clumps of cattails whose brown velvet harpoons might fancifully be described as mounting guard over the rejuvenated land. The near-death of this sea marsh was, as I have described previously, in part my fault. But its very slow renaissance is also attributable to the problems of cleansing the soil of poisonous materials accumulated through spraying and the run off from overuse of ground fertilizers, for a town storm drain also empties into this area.

But, inevitably in the natural course of events, it is not only the attractive and benevolent plants and insects that have made a comeback; the problem children (to us) of the natural world have also benefited. I have never seen so many or such large hornet and wasp nests as have appeared in the last few years, and this season there seems to be one in every clipped hedge.

Some of my neighbors are having their gardens devastated by a tremendous outbreak of a parasitic vining pest called dodder. This is exceptionally hard to eradicate; it starts as a rooted groundling, but once it starts to climb, the ground roots die and stem roots appear that grow into the tissue of the host plant. These steal the victim's lifeblood while strangling it in a viselike embrace. Clearly, dodder also has taken advantage of the moratorium on overhead spraying, and since the spores are airborne, I am on a constant alert. I have no wish to flame-gun my soil—the only nonpoisonous remedy if dodder gains a toehold.

But the worst plagues are, of course, the mosquitoes, which exist in larger and more bloodthirsty hordes than ever before. So bad have these become that a local poll is being conducted to see whether the residents of this town want to return to overhead spraying. I for one am deeply concerned lest the spraying be renewed, and I am grateful that we have been given a chance to express our views. Spraying may temporarily lessen the nuisance of mosquitoes, but it cannot eliminate the problem. Mosquitoes thrive wherever there are salt marshes, and they always will. What's more, those that survive an aerial spray pass on their newfound tolerance to their offspring, so that in a real medical emergency even more potent sprays would have to be used. In the case of dangerous epidemics spread by mosquitoes, I would of course accept spraying without question; what I am against is unnecessary use of powerful materials in order that we may have our cookouts in greater comfort. Perhaps we should try using harmless mosquito repellents on our arms and legs. What actually is needed is a much better educational program on the control and eradication of the areas where mosquitoes breed, which can even be a saucer forgotten among your flower pots with a little water left in it.

Nature is a matter of balance, and every living thing plays a specific part in maintaining this delicate equilibrium. We must learn to share the living world, not try to dominate it or twist it to our personal ends. If we are shortsighted and do not learn this lesson soon, we run the risk of handing on a deprived environment to those who come after us.

The year following the writing of this piece was extremely wet. June broke all records for rain and lack of sun. The mosquito population was therefore worse than it has ever been before. Unfortunately, this time the town did not consult the residents. Overhead spraying took place, without warning, very early one morning, even though we had been assured in our local paper that the fogging ground control method would be tried first.

It is now weeks since the plane crisscrossed this garden raining down poison, and it is still impossible to walk outdoors after sunset without being tormented by mosquitoes. And for the first time in four years I have not yet seen a single monarch butterfly—

Loosestrife

One of the few pleasures that can be gained from fighting the traffic at this season is the sheets of purple flowers stretching for miles wherever there are natural wetlands alongside the highways or swampy land formed by building an elevated road. The color extravaganza comes from innumerable small flowers that cluster around the tall stems of the common loosestrife, whose technical name is *Lythrum salicaria;* which is poetically described by a seventeenth-century herbalist as having "spiked heads of flowers branching out like spikes of lavender."

Loosestrife delights in wetlands and spreads like wildfire wherever it gains a toehold. It is not native to this country, but no one seems to know whether it was imported for garden purposes or hitchhiked across the Atlantic. It loves its adopted country, and it gives a great many people huge pleasure to watch the wind ripple over the tens of thousands of flower heads. There also exist garden varieties of this plant that come in a stronger red without the magenta tinge. Not unnaturally, considering their ancestry, these too grow better in damp places, but they do surprisingly well in the ordinary flower border with minimal care.

The purple variety that we now call the wild loosestrife is exceptionally tenacious and will, rather endearingly, hang on around the verges of wetland that has been reclaimed, long after every other sign of the original marsh is gone. If you are house-hunting at this season, it is wise to look carefully around the edges of any new development to see if you can spot patches of purple loosestrife in the vicinity. If these plants exist, you can be sure that the land is either part of the flood plain of some nearby river—with all that that name implies—or that it is filled-in marshland. In either case the chances are strong that the cellars in houses built on this type of land are damp and liable to flood. We should never forget that the flood plain adjoining any river is the natural overflow area from normal spring or abnormal freak floods, and that marshland became that way because it served as a sponge to collect surplus neighborhood water. Unless an enormously complicated underground drainage system has been installed—which, even if it exists, is not always effective—flood plains and old marshland will continue to fulfill their primary function, no matter how expensive or elaborate the houses that have been built over them. Purple loosestrife still growing doggedly in the vicinity is a great giveaway plant, particularly if a developer has been less than frank about the original nature of the land on which the houses have been built.

Although we enjoy loosestrife along the roads, it brings with it considerable disadvantages. The seeds are windblown and can carry a tremendous distance, and the plant is deeply rooted and very smothering to everything nearby in its habit of growth, for loosestrife ruthlessly chokes out local wetland flora. Along the eastern seaboard the biennial fringed gentian is an example of a plant that is in considerable danger of extinction in the wild because of the advance of purple loosestrife. In a sense, love it as we may, loosestrife is almost as much of a menace to our native plants as the encroaching bulldozer. For this reason I would be extremely careful about introducing it into a swampy place in any yard, and I have kept the improved versions that I do use confined to a hot, dry flower garden where the seed does not germinate.

But despite these precautions, the purple variety has recently appeared in a few isolated clumps along the edges of our saltwater marsh and gives the impression of happily and rapidly starting to make itself at home. Its route is blocked seaward, for it dislikes brackish water, but I notice that it has already spread landward and is starting to jostle some of the pampas grass. Now I shall have to try to dig out the deep roots, something I ought to have done the moment I noticed the first plant, and I probably have several years of hard work ahead of me. But since this plant will not live and let live, I have no alternative, for there are local plants in our tiny reserve that I do not mean to see menaced by the purple plant eater!

August

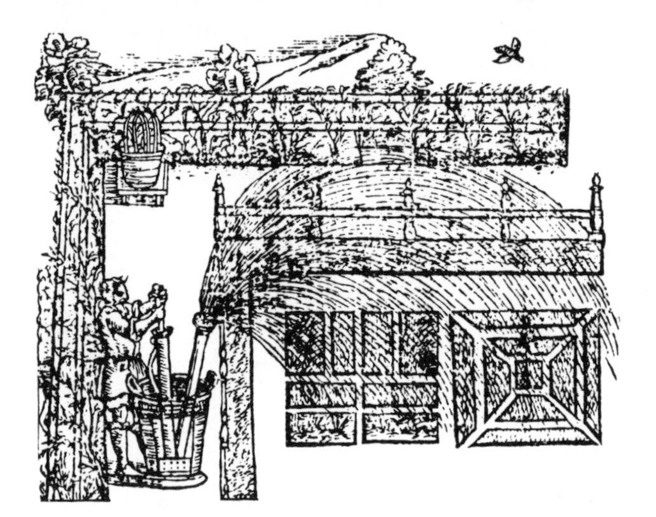

Summer Wild Flowers

Riding the miniature railway that winds across several miles of cranberry bogs in our area is a grandmotherly activity that is part of the ritual of summer. The excursion, which always takes much longer than planned, is extremely tiring and very noisy. Also, mysteriously, it usually takes place on the hottest day of the summer. Nevertheless, I always enjoy it, partly of course for the pleasure it gives the children, but also on account of the wonderful birds, butterflies, and flowers that can be seen in the remoter stretches of the bogs.

As the little train clanks along enveloped in a strong smell of hot popcorn, I try to avoid the closed-in carriages and settle myself onto a wooden bench in an open car. From this vantage point I can see all around me flooded land that has evolved, over the years that I have been looking at it, into deep black pools covered with thick sheets of water lilies, all crowded together so thickly that the stalky white flowers look like ballet dancers on a moss-green stage with the skimming swallows for partners. This artificially flooded land still has a great many of the old water-killed trees standing in it; with their leafless, gaunt black branches angled by the wind into grotesque attitudes, they have an unearthly look and remind me of old illustrations from the Grimm Brothers' fairy stories. And when, as sometimes happens, black and white egrets can be seen perching on those devastated branches, there is a slightly magical feeling in the air in spite of the prosaic reality of the train, the noise of the children, and all the panoply of tourists.

In the farthest reaches of the cranberry bogs there are high banks to contain the water when the vines are submerged. On these banks the wild flowers grow in extraordinary abundance. The chicory is taller and far more brightly blue than the specimens we see along the roads; the black-eyed Susans and the joe-pye weed are twice their normal size, and the wild asters and goldenrod are covered with clouds of butterflies that remain fluttering around the plants even when these are swayed to and fro by the air rush of the passing train.

These wild flowers of late summer are plants that we consider nothing but pleasant weeds. But in a curious twist, while our countryside has been swamped with foreign flowering weeds that we enjoy but, with the exception of the day lily, do nothing much about, our fall wild flowers have been turned into excellent border plants by European hybridizers and are used extensively to brighten up a slightly off period in the flower garden.

Goldenrod is an outstanding example of a plant that is misunderstood and a prophet without honor. Here goldenrod has been unfairly condemned as allergy-inducing, an evil reputation it seems unable to shake off although it is entirely innocent. It may have got this bad name because it flowers at the same time and often grows in association with ragweed, the classic allergy offender. And far from always having been an outcast, imported dried flowers of goldenrod were once held in tremendous esteem in Europe as the source of a healing balm for wounds. Alice Coats, in *Flowers and Their History,* quotes the seventeenth-century herbalist Gerard in his rather wry assessment of the fact that while goldenrod was worth an enormous amount when it had to be imported, as soon as it was found to have naturalized itself near London not only did the price fall spectacularly but the alleged medicinal worth of the plant also declined.

But although it may have lost its reputation as a healer, goldenrod remains a very beautiful plant that includes many different species. In dry areas it grows tall and thin, and the flowers swing in the wind like wheat stalks. On the banks of the bogs there are sturdy dwarf varieties with better leaves and a stronger-colored flower head. In brackish water on salt marshes a miniature variety can sometimes be found with flowers that are pale primrose color.

English gardeners are also extremely fond of a plant they call the Michaelmas daisy, a hybrid of many of our wild asters. During the European rage for bedding out, the style of gardening also known as *parterre,* in which all the plants used are annuals, not much was made of any of these American imports. But they came into favor and a great deal of breeding was done with them when the naturalistic herbaceous border became popular early in the twentieth century. Improved forms of asters and goldenrod are now being reintroduced into this country, while special Oregon-raised varieties of asters are now also available. If only for the sake of the butterflies they attract, these are worthwhile garden plants, but their culture, though simple, does take a little understanding. All of them—the improved black-eyed Susans (known horticulturally as rudbeckias), goldenrod, and asters—are gross feeders and take a great deal of fertility from the soil. A border containing any of these plants should be heavily top-dressed with coarse, half-made compost in the winter after the plants are cut down and kept under a thick mulch that steadily renews the soil all summer long. They also are tremendous spreaders when brought into cultivation. Most books suggest digging and dividing the plants in the fall, but in this climate that has not worked for me; new divisions do not take hold sufficiently, since the plant flowers too late to withstand winter. Asters should be divided in the spring. I pull out by hand at least half of every clump in early May, which produces strong, stout stems that are less crowded. In early July I shear off the top four inches of the tall varieties straight across if I want even flowering, and I layer the cuts at different levels if I want a pillar of bloom. This treatment delays bud formation a little, but asters are such sturdy growers that the setback is very slight. I do not shear the low-growing asters, but I keep after the spreading stolons and pull them out during the entire growing season. After flowering, the dead blossom heads are cut off, but the foliage is left to ripen and cut down only after it has been killed by hard frost. The fastest way to allow a border to go to ruin is to fail to cut down tall stalks of withered plants in the early winter. Left alone, these are used as a wintering-over area for pests; and weeds which get deep into the heart of the plant next spring before you get around to cutting the dead down will afterward prove unextractible. It is, however, a good idea to leave four-inch stubs of the dead stems; these seem to help the delicate varieties winter over better.

Asters and goldenrod are extremely prone to mildew wherever they grow; it is important never to water the foliage on humid evenings, for this will invariably start up an attack. The July shearing usually makes the tall varieties strong enough to need no staking. If you have to stake, do avoid the witch's broom effect. These are plants that are supposed to spread and flop a little; they do not look their best primly rigid.

I have found these plants a great pleasure in my yard, and I should be sorry to go into this time of year without knowing that their strong colors were on the way. But their usefulness as garden flowers does nothing to deter my delight in them in the wild when I see them in natural unimproved glory jostling each other beside russet-colored cranberry vines.

Around here we are extremely proud of our fall foliage colors; they have in fact become a considerable tourist attraction. But a lot of people also take their holidays in August, and I think we should do more to publicize the astonishing glories of the wild flowers of high summer. There is nothing like them anywhere else, but we take them too much for granted to advertise them as we should.

Not with Me!

There's a theory around that plants do better when they are well loved and constantly talked to by their owners. It also seems to have been shown by scientific means that they cringe quite literally not only when they are hurt but if someone even thinks violence in connection with them, and that they cannot thrive in an atmosphere of hostility.

I am inclined to agree that well-loved plants respond, possibly because they get more attention and small ailments are immediately dealt with, but I also have flourishing proof that hard thoughts, followed by drastic action, do not turn off all plants.

Mid-August is rather a standstill period in gardens. The turn of the growing year has come, and most trees, vines, and shrubs as well as perennials begin to gird themselves for winter. They stop growing and concentrate instead upon setting seed or hardening off the growth made during the current year. This may not seem obvious to gardeners still battling encroaching purslane and crab grass in the flower borders, while also cutting lawns and trimming hedges. But the autumn slowdown has in fact begun, and by late August, for instance, it is possible to move evergreens without any damage, for they are already semidormant.

But in the country garden there is a vine that refuses to conform to any conventional behavior. Nothing stops it from growing: neither the natural slowdown of nature nor the very hard thoughts loudly expressed about it in its presence by me! The offender is an old wisteria that grows to one side of our old-fashioned porches. When we first saw it the vine had reached to the top of the chimney of this three-story shingled house. In its hand-over-fist progress upward, it had totally destroyed two old-style wooden gutters which had been pried from the house by the thrusting suckers, had dislodged innumerable wooden shingles, and had edged itself through a broken screen inside the covered porch where it was growing away with all the enthusiasm of a vine in a greenhouse.

We rented the house for a season before we bought it, and while we were only tenants, there was nothing I could do about the wisteria except cut it out of the porch, for it was a haven for spiders. During this mild pruning I noticed that in spite of all the monstrous growth there were no signs of any seed pods, and the vine clearly had not flowered in recent years. When we bought the house some sort of control over it had to be instituted at once. Since I had had no experience with wisteria, I read up on it rather carefully, for I wanted to make sure that I did it no harm and that I redressed whatever ill treatment had prevented it from flowering in the past.

Book in hand, I pruned it as instructed at the proper season and rearranged the shoots to form an attractive canopy on a wooden lattice we put up specially for it. I then sat back to enjoy sheets of bloom—but I am still waiting and think that twenty-five years is a long wait without the slightest hint of blossom! After the first couple of barren years I tried in succession every possible trick suggested by the experts, and a few invented out of my slightly sadistic reaction to the plant's lack of cooperation. I have let it grow, as far as feasible, while still keeping it within bounds, cut it back, cut it to the ground, ringed the main trunk, root-pruned it, starved it, and fed it—all to no avail. Yet in spite of my strong feelings about it, that plant has never had an unhealthy moment or harbored any disease. Eventually I did some research about the lack of flowers. I learned that a phenomenon of this sort is not uncommon and is thought to occur either when the original cutting from which the plant was formed was taken from a side vegetative shoot that in itself could never bear flowers or when the plant had been grown from seed. In such cases there is nothing to be done.

When I happened upon this piece of information the solution seemed obvious: since I was never going to get lovely purple streamers from this wisteria, why keep it? Why not grub it out and replace it with a flowering specimen? And this is what we did, although it proved an immense job. Taking down all the foliage and getting out the roots took two men nearly a week. But eventually the old monster was dug up, the debris was carted off to the dump, and I set to work to make a new planting area for better wisteria. Since we get down here a little late in the spring, we decided upon the variety *Wisteria Issai,* which flowers slightly later in the season. Both my husband and I had had enough of struggling with the virility of the old vine, so we took special pains to procure a flowering specimen that was specifically described as being easily controlled.

But that poor little plant, and several others that I tried subsequently, never had a chance. Like rabbits in the grip of a boa constrictor, they were all strangled to death by suckers from the old vine—that sprang up like giants refreshed from minute sections of the old roots that had been overlooked. We dug over that patch of ground twice, taking out every strand of root we could see while removing the old soil. I also took care to set out the new plants much farther from the trellis where the old monster had grown. But it was all in vain; suckers continued to appear, and it was almost uncanny to watch the way in which they thrust themselves out of the ground alongside my better variety and scrambled up to the light (killing the new plant as they did so) by means of the stake I had set into the ground to help the replacement. No matter how carefully I untangled the intruders and yanked them out by the roots their presence was somehow injurious to the better vines, which inevitably succumbed during their first winter even after I had successfully nursed them through the summer. Eventually, of course, I gave up the whole experiment in despair and allowed the original wisteria to take up official residence again. And this it has done with undiminished energy, apparently refreshed by the struggle it had been through —but still resolutely refusing to flower. Then by chance last spring I went to see a friend who owns a large piece of property on which there are some enormous trees. As we sat on her terrace looking across at the woods some distance away, my eye was caught by a purple haze that looked rather like a waterfall streaming down from the top of a magnificent elm. It proved on closer inspection to be huge sheets of wisterias in full flower that had grown to a greater height than I had ever seen before. My hostess told me that she and her husband had set the wisteria out in the woods about thirty years ago, the plant having been dug up from beside their house, where it had never flowered. In the intervening years they had forgotten all about it, while presumably it was almost invisibly green, snaking its way up the tree. But six years ago, when it had reached an amazing height, it had begun to put on this tremendous show which was better each season. Her story was all too familiar; her vine sounded exactly like ours but with a much happier ending. And it now seems possible that apart from the problems of vegetative reproduction causing no flowers, there may also be an unrecognized species of wisteria on the market that will not flower until it has climbed like Jack's beanstalk and reached a certain stage of maturity. Many forms of ivy behave in an identical fashion, refusing ever to flower until they are old enough to have climbed to a considerable height and have also undergone a leaf change that appears at the flowering stage. I would have liked to have discovered whether that magnificent blossoming wisteria also had a slightly different leaf form, but the place where the long racemes of color began was so high in the tree that without binoculars I could not make judgment.

I cannot give our wisteria its head in this way. It would again ruin the house, and as I know from past experience, even when it has reached the tallest part of this building that is still not high enough for it to flower. But along the lines of what might have been, wouldn't it have been lovely if it had been set out where it could have climbed into the enormous copper beech that dominates our backyard? What a sight sheets of purple flowers would have made tumbling down against the bright red leaves of the unfolding tree.

Since none of this can happen, the only advantage to be gained from my experience is a warning to any possible purchaser of a wisteria that it is important to make sure that the plant has already flowered. And August will have to remain the month in which I settle down to the annual long-drawn-out struggle to pull this invasive vine out from under the shingles and off the downspouts where it still continues to thrive in spite of my irritation with it.

My Word!

There's nothing like the impending visit of a critical gardening relative for forcing the cold, hard horticultural deficiencies of your own yard upon you. In fact, one of the most ego-deflating activities any gardener can undergo is to walk into his own yard and look at it as though seeing it for the first time: it is then that all the weaknesses—the unshorn hedges, unweeded walks, sprawling plants, and flourishing crab grass—suddenly burst upon you in a new and most depressing light, and you wonder how they escaped your notice before and what on earth you are going to do about them.

I feel this situation particularly at the moment because a week or so ago I made just such a dispassionate inspection of our yard with very depressing results. And I am now hard at work trying to get the flowering area, at the very least, slightly more presentable in this lamentable season and ready for another in-depth inspection by my brother. There is too much that needs to be done in this small area alone for any rescue work to be undertaken in the rest of the yard, so other than a violent application of salt and boiling water in any position where ground elder has burst into public view, and the watering in of a non-soil-poisoning weedkiller onto the bricks, everything else will have to be accepted as it stands.

In a way this serves me right and is a warning that all gardeners should heed when they visit someone else's garden: Do not offer advice, no matter how eagerly it may appear to be sought, for it will come back to haunt you. Last fall I, unfortunately, did offer my brother a good deal of what I then considered quite sound advice about his flower garden, but in comparison with the condition of mine at present, his was paradise regained, so I am feeling extremely foolish!

I have now fed and sheared back for the third time what should be a solid ribbon of white petunias that edge both sides of the central path in that garden. But instead of a solid unbroken mass of white flowers, all I have is a tattered slimy mess. Behind the petunias the rows of zinnias and snapdragons have been flattened by the incessant torrential rain and torn apart by the high winds that have been such a feature of the past weeks. And even if I can stake them upright again into something approximating their original position, the end result will be far more stake than flowers.

Nature has not confined itself to making trouble with rain and wind; it has added the complication of unusual activity on the part of nectar-gathering insects, for they too have been suffering from the bad weather. A huge, self-seeded white veronica stretching back into the groups of platycodons has sprung up from nowhere among the petunias alongside the path. Veronica is a coarse plant and not particularly exciting, but it does have the advantage of being very full of bloom, and the extraordinary disadvantage that the flowers seem to send insects crazy. Every spike is quivering with the weight of bees and enormous wasp-waisted black monsters which are built along the lines of huge hornets only much larger. The bees pay no attention to me as I pass by, but the black monsters rise with an angry buzz, being apparently both sensitive and very nervous, and hover around like tiny helicopters whenever I go near the petunias. I have never seen these particular black giant hornets on anything except allium blossoms before, and I am frankly terrified of being stung by one of them, since I am fearfully allergic to bee and wasp venom. This hazard prevents me from tidying up the area in the immediate vicinity and taking the dead flowers off the platycodons. And if I don't get to the latter job, the plants will cease to bloom, removing one of the only good color shows in the garden.

Although I can't do anything about it, I am rather intrigued by the behavior of the hornets and bees around that veronica, for the plant has, as far as I can judge, a twin, also self-seeded, on the far side of the same growing area that never seems to attract any insects. There are plenty of bees working over the smashed zinnias immediately around the neglected veronica, so it cannot have gone unnoticed; yet its flower spikes remain vacant while scores of bees and hornets struggle for room on the other one. I dare not put my nose down to the popular plant to see whether there is a difference in fragrance, but it would be interesting to see if that is the reason why one is the belle of the ball and the other the unloved wallflower.

In spite of their need of steady moisture, which has certainly not been lacking, this has also been a bad year for the phlox. The heavy flowering panicles of the new varieties seem overweight on thin stalks, and they carry so many flowers crammed closely together that with the constant wet weather the center of the influorescence rots. But even the older, lighter-trussed varieties have not done much better, presumably because of lack of sun. For the same reason the cosmos and African marigolds are all foliage with very few flowers, and the Boston daisies, normally the simplest of plants, have also run all to leaves with not a bud in sight. In this case I think the rich soil that has been built up by the regular application of compost may be the cause, for Boston daisies or marguerites flower best when they are slightly potbound or in poor soil, and the growing area bears not the slightest approximation to either of these conditions! As for the more delicate or slightly unusual plants I try to grow, they are all long gone to the compost pile, for those that were not spoiled by slugs or borers collapsed under the combined onslaught of mildew and mold brought on by the damp weather and the high humidity. The general effect of the entire growing area is a combination of a battlefield and a dump. I can see no way to bluff it out, for nothing else in the garden has done sufficiently well for me to be able to divert my brother's attention.

What is needed is some natural disaster that would take the burden of this badly grown area off my shoulders. No one who has lived through anything like a hurricane would ever want to go through anything of the same sort for purely selfish reasons, but I could do with a small, very discriminating catastrophe, something like a tiny tornado. Ideally this would leave my neighbors' gardens entirely unscathed, steer clear of the flowers on my porch and under my lath house, and spend its entire force devastating the flower garden without of course harming the hedges that surround it!

That way I should get sympathy and understanding instead of what I know is coming: long and highly significant silences. For I don't have to contend only with the forces of nature; there is a more subtle battle to be won. It is not easy to convince anyone who gardens in the British Isles that wind, rain, and lack of sun are sufficient cause not to have a first-class flower garden.

The Three Sisters

A neighbor who is an ardent horticulturist has introduced me to some delightful nostalgic experiments in raising vegetables that appear to combine all the virtues, for they are pleasant to look at, easy to carry out, and extremely productive. The aim has been to raise the three staples of the local Indian diet—corn, squash, and beans—all together on a small plot of land in a way that they might have been cultivated by the Indians before the arrival of the white man with his more sophisticated tools.

These experiments were in part an amusing hobby, an attempt to see if the old methods still worked, but there was also a serious side. For if they proved successful they might make it possible to grow a great deal more home produce on one small patch of land, which could be very important to people who are anxious to grow a variety of vegetables in a limited space.

The area I saw in action was a small plot in full sun with good air circulation that normally might be given over to a fair-size stand of sweet corn as the single crop. The land had been grassed over for some years, and when it was opened up, it was heavily fertilized with organic products that included dry cow manure, wood ashes, and some lime. This was raked well into the soil but not dug down deeply, for the Indian squaws were able to till the land only shallowly with hoes made of the shoulderblade of a deer tied to a stick.

As every child knows, the Indians used both fish heads and whole fish as fertilizers buried alongside the corn to give the searching roots the essential elements of nitrogens, phosphorus, and potash. Most of us admire the ingenuity and knowledge that led to this form of manuring and assume that was all the Indians did about keeping up the fertility of the land. But they did a great deal more for the soil structure than just burying those fragrant fish. The Algonquins, Iroquois, and other Indians also set aside special areas in which they piled up dumps of wood ashes from their fires, together with offal and other kinds of organic trash, which were eventually spread on the land. It does not seem likely that the dumps were left intact long enough to turn the refuse into what we would call worked compost; more probably it was spread over the area under cultivation in a half-decomposed state—a method we call sheet composting, in which the final breakdown of the waste products is allowed to take place on top of the soil. This may have been done in the fall after the crops were harvested. Presumably the Indians did not mind the smell, which must have been extremely pungent, and presumably bears and raccoons were kept from rummaging through it by the boys who mounted guard day and night over the crops and over the cultivated land on high wooden platforms built in the fields. But no matter how strongly it smelled, such a mixture must have added immensely to the continued fertility of the land—something that was absolutely essential if the Indian method of growing several crops simultaneously in small areas was to succeed.

It would be extremely hard to get away with anything like that Indian mixture for renovating soil today: neighbors have been known to complain even about the smell of decaying grass! But if land is to be used for multipurpose cropping in the way my friend was demonstrating, fertility must be kept enormously high. One way to do this is by spring fertilizing, and then, after the land is bare in the fall, a slightly more acceptable method of sheet composting can be used. This involves spreading hay or grass clippings or even fallen leaves thickly as a mulch over the cleaned-off ground, and then adding either a mixture of Bovung and bonemeal or else Milorganite an inch deep to cover the mulch. This will speed up the decomposition of the mulch material and also provide extra fertility to the land. A really thick winter mulch also preserves the quality of the ground underneath it—the goodness does not get leached out, and the earth does not pack down hard. When spring planting time rolls around, there is no need for spades or rototillers, for earth protected by a winter mulch emerges in a lovely workable condition—something the Indians knew long, long ago.

In the experimental plot, corn was sown around the end of May using the old-fashioned "hill" method—planting the seeds into earth slightly mounded up for better drainage. The mounds were set closer than normal, about three feet apart—a great advantage on a small plot. When the freshly sprouted plants were three inches high, they were thinned to three stalks per hill. Squash seed was then planted to one side of the little stands of corn, and the late-climbing Kentucky Wonder beans to the other.

The germination of the beans was good but slow, for they need ground that is thoroughly warmed up. The squash did not do so well, but this hardly mattered, for enough appeared to form a rapidly spreading carpet that soon covered the ground. All squash have large leaves that form a dense moisture-retaining mat. These deterred weeds by shading the ground, and also kept the corn and bean roots cool when the fierce summer heat set in.

With the excellent soil conditions, corn and beans grew strongly, the corn moving ahead the fastest and forming stout stems up which the beans slowly twined. When it seemed necessary to keep the beans from taking over a corn stalk too fast, the growing points were pinched out.

The whole project was a great success. The climbing beans did not hinder the wind pollination of the corn, and the ears received enough sunlight through the vines to ripen properly. The squash rather overdid itself, turning into a rampaging jungle that made it hard to spot the fruit while it was still an edible size, but all three species cropped heavily. Above all, the method eliminated the tiresome necessity of providing stakes for climbing beans. So this multiple use of a small plot proved well worth the slight extra amount of work that had to go into the preparation.

The Indians used corn, beans, and squash as dried staples in their diets, and it did not matter to them in what order the crops ripened; all they needed was plenty of everything dry on the vine. They harvested these crops when everything was fully mature, probably early in October. Modern horticulturists trying to get the best out of such a patch could take a slightly different approach. It would, for instance, probably be wise to use an early-maturing variety of hybrid corn—one of the improved kinds we owe to the late Vice President Henry A. Wallace—for the silk should have a chance to bloom and the ears plump up and be reachable for harvesting before they are overwhelmed by the climbing beans. Kentucky Wonders are delicious, tasty beans, but they are extremely temperamental and often don't do at all well. A more reliable choice might be one of the new stringless scarlet runner beans marketed by all English seedsmen and available here as a flowering plant. Scarlet runner beans are a great delicacy in Europe; they begin to bear after the bush beans are finished, and I regret that they are so rare here as a vegetable, for they are easy to grow. If they ever come back in favor in this country there would be a certain irony in the situation. For these are native American beans that were introduced into Europe in 1633 from the New World; this country has forgotten them, while they have been gaining ever increasing acceptance as a gourmet's delight in Europe.

Scarlet runners, as the name implies, carry masses of scentless red flowers that look exactly like a sweet pea. But don't use the ornamental types offered for sale in hardware stores for your climbing beans, for these are unimproved types and can be extremely stringy. Send away to one of the big European seed houses, Suttons of Reading or Dobbie of Edinburgh, and buy from their catalogs. If you are going to take so much trouble with your multiple-purpose plot, this extra effort is worthwhile. And you will get long green edible pods dripping like rain from the corn stalks, as well as flowers for aesthetic delight.

The term "hill" which I used earlier to describe a method of raising the ground slightly for better drainage is still in common use, although the practice of mounding up the area where corn is to be planted has been abandoned. Modern custom suggests that corn be planted in furrows—or, where rainfall is short, even in slightly depressed areas, to catch every spare drop. This is fine for rainless parts of the country, but it is not good practice where there is plenty of summer rain in a normal year or where the soil is heavy. Most vegetables grow better in moderately raised beds, even if this calls for occasional extra watering, for raised soil always remains well drained and is therefore much less liable to pack when it is constantly walked over to harvest the crops. In the experimental plot the heavy shading of the squash leaves made drying out a minimal problem even during an exceptionally dry summer. The following year, when there was an abnormally large rainfall, only crops planted in raised beds came through—the rest were all washed out in the unending downpours. So whether you are growing corn Indian fashion or just running a small vegetable plot, slightly raised planting here in the Northeast makes for better success.

There are variations to this Indian method which could also

be used by gardeners trying to get more out of their land. If the ground is kept under a foot of mulch all year round, it is possible to plant corn seed nine to twelve inches apart—not in the raised furrows—with a space of two feet of hay-covered ground between each row. Beans can then be dropped in alongside the corn seven to ten days later, eighteen to twenty-four inches apart. In this case the rampaging squash can be omitted, for the thick hay mulch keeps the ground cool and weed-free. In soil perpetually under mulch, seed can often be planted *in* the mulch itself if this seems simpler—some organic gardeners are in fact getting much larger returns on potato and other crops planted on top of a hay mulch—and then very lightly covered with more of the same material than from soil planting. And there is no comparison with the cleanliness of the crops raised this way and the simplicity of harvesting—something that is important to remember for those who are not up to wielding a spade.

So in spite of all the ifs and ands, and with a few modifications, the old Indian method remains viable and full of promise for the small landowner. I am not one who believes that everything old was better and that everything new is bad; what is important in this case is that the old method offers exciting possibilities to new gardeners with enough curiosity to take a little extra trouble.

For the details of this experimental work I owe warm thanks to Mr. Albert S. Bigelow, author, athlete, artist, and sailor, as well as a superb horticulturist, someone who cares deeply about finding better ways to use and preserve our land. The knowledge and skills were all his, but any errors in the deductions must be laid at my door.

Stripes

My grandfather was a clergyman, so the day always began with family prayers held in the dining room that overlooked the garden. Grandpapa was inclined toward long meditations, hard on a restless kneeling child, and to save me from his subsequent wrath, for nemesis fell swiftly upon children then, I used to be stationed near the window so that I could look out and thus while away the time. On the whole this plan worked; I can recall counting the number of red roses I could see during what must have been a very long exhortation, for the rose garden which was immediately below that window was extremely large. But occasionally the idea misfired. If I saw the stable pony hitched to the cumbersome lawn mower I fidgeted worse than usual, because I longed to get outside to the delights that lay ahead.

The pony was mainly used to pull carts of manure and swinging tanks of water around the garden, but although I loved him dearly I was strictly forbidden to accompany him on these duties. Apparently I had a limitless capacity for coming in either soaking wet or covered with dirt. But I was allowed to help him mow the grass, and my self-imposed task was to trot beside his head keeping the flies off him with a special whisk someone must have made for me. In hot weather—hot, that is, by English standards, possibly a torrid 70 degrees—the pony wore a straw hat with holes for his ears, through which they twitched and flickered. In the heat, his face was also draped with a blue beaded string net to keep the flies out of his eyes and off his muzzle. And no matter what the weather, whenever he cut the grass he wore special flat leather boots to prevent his hooves from ruining the turf.

I took on the job of keeping off flies with enthusiasm, and when he and I and the gardener, to whom I must have been a great nuisance, went out to mow, I always picked small bunches of wild flowers—daisies, buttercups, or dandelions—and bunched them together with a wiry stem of grass. One went to the gardener, who stuck it in the buckle of his apron, for all gardeners wore long blue aprons in those days; one was stuck into the tattered brim of the pony's bonnet; and one I sported in the broad black ribbon of the sailor hat I was compelled to wear outdoors. And thus bedecked, the three of us paced slowly to and fro across the grass, making those lovely stripes that are the hallmark of a well-kept English lawn. I was never happier than when helping mow the grass, and to this day the smell of a fresh-cut lawn (if you can get a whiff above the fumes of the gasoline used on modern mowers) brings back a feeling of security and contentment.

Today everything about that scene has changed. The slow-moving pony, the white-pinafored child, the somnolent clatter of the cutters blending with the cawing of the rooks in nearby trees belong to a moment out of an impressionistic past. The only thing we have today in common with that scene is lawns, which still demand mowing and which continue to be a status symbol for many homeowners. But grass in this country does not bring with it serenity and contentment; on the contrary, it is a hard master. If you take it too seriously it can spoil the pleasure of a yard by enslaving you, and the ravages of other people's dogs on lawns have made for more trouble than the worst spite fences. I have long given up being owned by my grass. I sometimes regret its appearance, but I have accepted the fact that without far more labor than I can afford or spare time for myself, I must make the best of what I have, so I have abandoned the grail-like search for velvety perfection. But even for those of us who have learned to be more casual about our grass, there are certain seasons when the way it is treated can make a great difference to both its immediate and its long-term health. One of those periods is right now in the heat of the summer.

When the sun blazes down, most grass stops growing and not much mowing needs to be done. But if you must cut grass in hot weather, always make sure that the cutting blade is set to its highest point. Grass roots need shade from hot sun, and leaving the lawn grass higher than usual provides shade. So this is the time of year to go longer between mowings, and when you cut, cut high.

Water is another problem. If you live where there are likely to be water restrictions it is wiser never to start watering. Grass accustomed to an artificial supply of moisture will die if water is cut off, and this can make even a short holiday a problem. If you live where grass must be watered to stay alive, plan to reduce the size of the lawns by adding ground covers along the margins to reduce the amount of grass. Water for gardening is going to be in increasingly short supply, and we should not plan a layout that demands care we may not be able to provide, for that's the direct route to frustration.

When and if you water your grass, do so on a regular schedule and be sure that the extra water penetrates deeply. Put out a flat pan so that the water from the sprinkler collects in it. The extra water will not have penetrated deeply enough to help the roots of the grass unless an inch or more of water has accumulated in the pan. Daily light sprinkling may keep grass looking greener, but it is not good for the long-term health of the lawn. It brings grass roots up to the soil surface and exposes them to the scorching effect of hot sun.

Light watering also encourages the germination and continued survival of crab grass. This shallow-rooted nuisance will not thrive nearly so well in lawns that have regular deep watering every other week. In general, benign neglect is the best treatment for lawns during the dog days, for then the grass will fall dormant and ride out the heat looking rather scruffy but keeping the roots safely alive in heat-induced light sleep. Later, when the weather cools off, the grass will start up again and there will be plenty to be done. But for the moment, let it be.

Tuberous Begonias

August is the season when the tuberous begonias reach their peak, and these are plants that separate the men from the boys where horticultural skills are involved. Most tuberous begonias plunked into pots will send up some kind of flowering stem as long as they are not grown in full sun, but there is a world of difference not only in the appearance of a properly grown plant but also in the length of time in which it can give us pleasure, and since the tubers are quite expensive a little more trouble is worthwhile.

I am a casual gardener, slightly inclined to feel that a plant must put up with my ways or I am not going to bother with it. I have always grown moderately successful tuberous begonias that were fine as long as no one looked at them too closely. But in recent years I have been visiting a friend who grows the most magnificent specimens of these plants, and the long-lasting pleasure that she got from less expensive tubers than those I had bought made me decide to take stock of my methods and try to follow expert advice to the letter. This all started two or three years back, and though I am still inclined to hide my plants when my friend visits me, I have done far better than in the past; and there is no question that I am getting much more for my money.

With tuberous begonias the story has to start very early in the gardening year, long before most of us are further on than dreaming about the summer garden. But when we leaf through those enticing catalogs there are often pictures of magnificent tuberous begonias and also shady flower beds in which they have been planted out to perfection. When we see those illustrations in the catalogs we should send in our orders right away. I think it is better, if possible, to buy the tubers through growers who specialize in these plants, for they will arrive in better condition. But if you never order by catalog, pick up the tubers just as soon as you find them offered for sale in your local garden center, which is usually sometime in March. Most of these tubers are imported and have already spent far too long in the plastic bag with a sprinkle of peat moss in which we find them, and the longer they remain out of the ground and in warm storage the more moisture they will lose. This, of course, shrinks the tuber, and a shriveled purchase will never produce the plant that comes from a nice fat cold-storage bulb. So buy these tubers the moment you spot them and plan to grow them from the very start for yourself.

This, I agree, is a good deal to ask of the average gardener, for it means at least three months of indoor care here in New England before the plants can go outside. But I have had such unending troubles with prestarted tuberous begonias picked up later in the season that I have come to the reluctant conclusion that there is no point in growing the plant at all unless you are into it from the very beginning. No plant suffers faster from rot in every area—whether the roots, the tuber itself, the stems, or the leaves—in a humid summer. Even the flowers get into the picture and rot too, and unless you buy prestarted plants from growers who make a specialty of begonias (and these people can't be found around every corner), most commercially prestarted plants are not worth the money, for they are usually set out in unsuitable soil without any internal drainage inside the pots, a double jeopardy that will inevitably bring disappointing results.

Not all tuberous begonias have to be those tall plants with rose-shaped flowers in every color except blue. There are also hanging begonias, which have slightly smaller flowers and, according to the pictures, fall in lovely cascades over the edge of the pot. I have never had much luck with hanging begonias, although I bought new stock several years in a row and tried them in all sorts of positions and with many different soil mixtures. They always turned into rather spindly plants with weak stems, and not too many of them. If you are a complete novice, I would advise steering clear of hanging begonias, at least at first.

Far the simplest to handle are the multiflora types which flower mainly in shades of yellow, orange, and red. These do not grow as tall as the ordinary tuberous begonia, but they put up a multitude of stalks and a great many leaves. The blooms are less spectacular but come in great quantity and last longer. This is the best kind for the beginner—and it is also the only sort to try outdoors, in spite of all those enticing advertisements, if you are a novice grower. I am extremely fond of multiflora begonias and not so entranced with the improvements made to them by the hybridizers that have resulted in a variety known as grandiflora, which come in the same color range but grow taller with slightly more brittle stems and can turn into something like oversized semperflorens begonias. But if you have cut your teeth successfully on multifloras, it is worth trying some grandifloras, for grown well in pots they can turn into pyramids of bloom.

As soon as you get hold of the tubers, start them into growth by putting them round side down into damp peat moss. Don't have the peat moss too wet; remember the deplorable tendency of the entire family to rot. I have seen it suggested that the plants be started in what is to be their final pots. This, I think, is a mistake; begonias ultimately have to go into very large pots. Starting a plant in a large pot containing masses of soil into which the roots are not ready to penetrate is inviting disaster. There is no need of any special equipment except a bag of dry peat moss, which can be moistened up with warm water and then wrung out like a sponge. It is easier not to use the conventional wooden or plastic flats, for these drip onto windowsills and offer later difficulties with transplanting. This is the moment for a little home recycling. Reseal the open end of a half-gallon milk container and then cut out an entire length of the long side; put in the peat moss, and you have an excellent starting box. No matter how often you have heard or read the advice, do not plunk the tuber down on top of the damp moss and allow it to root only from the bottom; bury it completely with a thin layer of peat moss over the hollow top. The reason for this is that roots grow from every portion of the tuber if it is covered. By cutting half the tuber off from the possibility of making roots, the large top-heavy plant that a tuberous begonia eventually forms will be deprived not only of useful anchor roots but also of much-needed nourishment. It should not be forgotten that all begonias are gross feeders and need all possible root action.

Put the planted package on a warm and sunny windowsill, and soon small pink sprouts will appear from the top of each tuber. Now there are several options open to you according to the kind of begonia you are growing and your personal wishes.

If you are raising multifloras or grandifloras, leave all the little shoots alone. If you are growing the more common big-flowered variety, you will have to decide whether you want a single-stemmed specimen with gigantic flowers or one that throws up several flowering stalks. Like most things in life there is something to be said on both sides. Plants grown to a single stem are far easier to stake, and since all their energy goes into one treelike trunk the flowers come in a steady procession. But—and it is a real but—those huge stems with the enormous leaves that will grow from them are knocked over easily by wind, and even if this is avoided, wind can tear the big leaves. All begonias are liable to stem rot, not always because they are badly grown, but sometimes because of the weather. If a single-stemmed plant collapses then there is nothing left of the show, which is not the case with a plant that has been allowed to put up several flowering stalks. I compromise. I allow the old tubers to carry several stems, but I rub off all but the strongest of the pink snouts in new tubers.

The peat moss should always be kept pleasantly damp but never wet, and the plants allowed to develop slowly in a warm, sunny window. I keep mine under artificial lights at this stage, and if you own such a unit it could not be better employed. At maturity all begonias do best in light shade with some sun during part of every day, so full spring sunlight at the window will not harm them. Keep the container regularly rotated. Tuberous begonias face in a single direction, as the leaves will soon show you. But you may not have planted the tubers so that these all develop in the same direction, so every side must have its turn with full light.

When the leaf buds begin to expand, add a little highly diluted liquid fertilizer to your mister and spray the leaves and the peat moss. As soon as any tuber develops two leaves the same size, it is ready for potting up.

For this there has to be a special soil mix, and because the tubers can go into anything up to eight-inch pots, mix up a large batch. Since I have been taking trouble with these plants, I have found that a rich compost with a quarter of a cup of perlite added to each cup of compost suits mine the best. I do not add peat moss (which I often see advocated), for the tubers are already rooted into that material. Every grower has his own soil formula, and the only essential factor is that the mix should be rich and well aerated—that is, fluffy and well enough drained so that surplus water runs through it fast. For the potting process there should be at least an inch of broken crockery at the base of each pot: I find such shards more satisfactory than pebbles, although the latter will do in a pinch. But it is my experience that pebbles pack down under the weight of the soil, while shards retain their structure and keep air space between them.

After the drainage material is in, add a couple of inches of soil mix, but don't pound or press this down. Then spread a half-inch of fish meal, which can be bought at most good garden stores, over this first layer and stir it into the soil with a stick. I have had some trouble with squirrels digging out my carefully planted tubers to get at the delights of the fish meal, and this is something to remember when the pots go outside. Place them out of the reach of any animal. Add enough soil so that when the rooted block of peat moss is in place the bottom of the flower stem will be about half an inch below the pot sides. This will allow ample space for watering.

If you have sprouted the tubers in milk cartons the next step is child's play. Take a sharp knife and cut through the carton walls between each tuber, then peel off the cardboard and plant the peat moss block. If the tubers are in a solid flat, score out the blocks in squares with a very sharp knife and pry one out gently with a blunt rounded knife. The rest will come out easily.

Once the root block is on the soil mix, pour more new soil in all around it to fill in every chink, but don't add any on top of the tuber, or you may rot the base of the stem which has been used to growing in the open. Settle the plant into place with some gentle thumps, but don't firm it in. I once heard a professional grower of begonias describe the whole family as "lazy plants," meaning that the roots dislike having to penetrate rammed soil. If you bear this in mind while transplanting you will have no problems.

At this stage with the large flower-bed begonia, put a stout stake deep into the pot behind the flower stem but not through the peat moss root block. Tie the stem to the stake with something soft; string will cut through the juicy stem. I use paper-covered wire, which seems to work well, but don't lash the poor thing up. Make a looped tie that will allow the stem to expand. As the plant develops there will have to be regular tying in, and this is one of the small tricks that produce better-looking plants. At first the pot will look all stake, but as you tie in the elongating stem you will be thankful it is so long and so stout—a small stake never did anything for a tall begonia. By regular staking you train the plant the way you want it to grow; late staking often gives the impression of a martyred victim of religious intolerance.

If you bought prepotted plants from a professional grower

who uses a soilless mix, they will still need stakes. You must also remember that plants in such a medium need to have a regular pinch of water-soluble fertilizer with each watering.

Some growers feed their plants with a weak dilution of fertilizer every second week. I do not go in for such heavy feeding; I find that my own rich soil, combined with the layer of fish meal, provides all the nourishment needed; more produces a flabby top-heavy plant. To judge, watch the leaves. If they curl under, the plant is overfed. If they brown at the edges, there is too much light and possibly too much water. If they look pale, they need a booster shot of fertilizer.

Always be extremely careful when watering that tree trunk of a stem, for it is dreadfully liable to rot. And the rot sets in if damp soil rests too long against it or if water is splashed carelessly around. There is no way to explain how to water carefully except to remind you that plants prone to mildew and rot like tuberous begonias should always be watered in the early morning so the foliage can dry off before night, and if possible use a watering can rather than a hose for better control.

Almost none of these problems arise with the multifloras—they are less liable to mildew, they don't need stakes, they hardly ever need extra food, and it is extremely rare for the stems to break off. The grandifloras do need small multiple stakes, for their stems are more brittle.

If the worst happens and the main stem snaps off, move the pot to a shady place out of the rain and continue moderate watering. Since the plant was cut off in its prime before the leaves had finished their work of sending nourishment back to the tuber, the chances are excellent that in a short time new shoots will appear from the decapitated tuber. These should be allowed to develop as long and as strongly as possible. If a pot that has lost its only stem is dried out immediately, the tuber will die as surely as though you threw it in the fire. I used to make this mistake, and sometimes the unhappy plants would try to send up new growth in midwinter in the cold dark cellar with fatal results.

If you have managed to keep the plants growing until frost hits, this will finish off the foliage, but since the roots will still be active, the pots should be put aside, or the plants left in the still-warm ground for at least a week. Drying out the pots involves slowly diminishing the supply of water, a process that should go on for a month. Then the pot soil can be allowed to become bone dry. With plants in the open ground, lift the tubers a week after the foliage has been killed, and ripen them with the soil still on their roots in a sunny but not too hot place. Never clean off roots or foliage; allow everything to wither slowly of its own accord, for this is the natural method of settling the tuber into happy dormancy. Remove debris only when it is entirely dry.

After all is quiet on the growing front, I rest my pots several ways. Some I put on their sides under the staging in the greenhouse where they get no moisture in the pot soil but live in a humid atmosphere. Some I leave bone dry in their pots in the cold cellar where the temperature never goes above 55 degrees, and some I take out of their pots or dig up, clean off, and store in dry peat moss in the same cold cellar. I have had fair results from all three methods. Each year I lose a few tubers from a sort of dry rot that does not disintegrate the bulb but solidifies it. On the whole the pots that have done best for me are those stored in the greenhouse after they have been given a chance to grow on until they demand to rest. If you do not have these facilities, just make sure that your tubers are stored in a place where they can be kept both dry and cool.

This may all seem, and indeed is, a great deal of trouble for a plant that can be made to flower without anything like this effort. I by no means do all the right things every year. But when I do take the extra trouble I have tuberous begonias that are light-years better than the more casually grown plants, so for the perfectionist these are the optimum rules.

September

Gloxinias

This is obviously going to be an excellent year for those members of the gesneriad family that flower outdoors in pots in the summer. I have a quite exceptional display on my porch from my old plants, of which there is now an immense collection, for I usually add a few new ones each season, and the show is a scene stealer by night when the display is illuminated.

Gloxinias (which technically should be called sinningias) have been rather turned back to front in our minds because the retailers now mainly offer them for sale in the spring. Many gardeners therefore think of this plant as being in the same category as a pan of spring bulbs, particularly since gloxinias grow from something that looks a little like a bulb, but is in fact a modified rhizome that we usually call a tuber. The sight of a dozen or more gloxinias in lavish bloom on a September porch is therefore regarded as something rather odd by many who pass by, whereas in fact the plants are doing what comes naturally by flowering at this season.

To get gloxinias to flower in the spring the growers force them with extra heat and humidity and many additional hours of strong artificial light in greenhouses. Gloxinias are willing to take this treatment and change their ancestral flowering habits—something many plants just will not do no matter how hard professional growers try to manipulate them. But the result of the artificial life style that brings them to bloom at the "wrong" season is that the plants can't make the change from the forcing house to our houses and usually collapse without opening all the enormous buds they carry. In consequence gloxinias are getting a bad name as difficult plants, which is a shame, for correctly handled they are easy to grow and extremely rewarding.

Grown naturally, gloxinias need the long daylight hours of summer to bring their huge foliage growth to maturity and set bud. So if you want a fine, long-lasting plant don't allow yourself to be charmed by one already in bloom next spring, but instead buy a gloxinia tuber that already has signs of tiny sprouts. Grow it slowly all summer, treating it in exactly the same way as a tuberous begonia. No one is scared of trying tuberous begonias—I see them everywhere—and yet many people are frightened of so-called difficult gloxinias. But with a few minor modifications, which I will mention later, there is no difference in the culture of the two plants, and with a fall-flowering gloxinia, you will be rewarded with a stupendous plant that won't faint and fail when it is brought indoors after the buds begin to open, which happens with tuberous begonias that cannot stand indoor life at all.

But eventually the gloxinia will have no more buds to open, and the leaves will begin to look untidy. Then comes the question: Should the plant be thrown away or can it be carried through the winter dried out like the tuberous begonias? Almost every book suggests withering off the foliage by a slow withholding of water and then storing the cleaned-off tuber either bone dry in its pot or else in dry peat moss. But here it is important to realize that although the plant can be treated while it is in active growth exactly like a tuberous begonia, it is not a member of the same family and does in fact do better with different winter handling. For years I followed the conventional advice to the letter and I could not fathom why most of the gloxinia tubers refused to sprout in the spring even though they looked fat, juicy, and extremely healthy. No matter how I handled them—and I tried every conceivable possibility from lights to heating cables to plastic bags—the majority of the old tubers sat there in a sort of horticultural limbo neither dying nor growing. Any tuber that had already started to grow gave me no trouble; it was the bald pates that I could not get going, something that also happened with unsprouted new tubers that I bought. Eventually, in disgust, I stopped attempting to save my old plants and made very sure that any tubers I purchased were already showing signs of growth, for, ask where I would, no one could explain my problem.

Then, by chance, a great expert on the various members of the gesneriad family, Mr. Michael Kartuz, of Wilmington, Massachusetts, was with me on TV. We planned to discuss African violets and their kin, and had laid out a program that fitted the time slot exactly. But toward the end, as the time cues began to go up, I realized with dismay that we had gone faster through our demonstration than had been expected: we were about to run out of anything to say or do with several minutes still to go. In desperation I snatched up a large flowering gloxinia that was there only to decorate the set, and thrust it upon my unsuspecting guest, asking how best to carry it through the winter. I planned to fill in the time by explaining my difficulties. But to my surprise I was offered a radically different approach to winter treatment of gloxinias, for the expert briskly twisted every scrap of bud and foliage off the unhappy plant and handed me back the pot with the poor scalped tuber showing above the pot soil, saying in effect that I should treat it like that, wait for it to resprout, which would happen fast, and then allow it to grow on all winter.

I went home, considerably bewildered, clutching the miserable remnants of that plant. The advice I had been given seemed to contradict everything that common sense would indicate to be feasible treatment of an exhausted tuber. But since my success rating with gloxinias and dormancy was in the zero percentile—I had borrowed that plant—it seemed I had no choice but to try out this remarkable approach with the gloxinia that had been so spectacularly denuded on air. And it all happened just as the expert had prophesied. The tuber resprouted in an astonishingly short time, although I made sure that the pot soil never got sodden while it was making this comeback, and I carried the plant successfully through the winter in a cool, bright, but sunless window. Eventually, rather thankfully, I was able to return it to the original owner (who had watched the show and was also more than a little dubious) full of bud.

Now I treat all gloxinias that way. As soon as they finish flowering, off comes the foliage, and the pots are set aside in a cool dry place with minimal watering until new growth shows. Usually the beheadings take place before we go back to town, for this makes the pots far easier to pack. If a plant still has some unopened buds when the crown is twisted off, it will open its flowers indoors if the cut stem and spreading foliage is balanced on something like a soup plate with water just touching the stem. Not all the pots come back at the same time, even though the foliage is removed simultaneously; there can be as much as a month's difference in the period between decapitation and regrowth. This is partly the reaction of slightly different varieties, but mainly comes from the maturity of the foliage. If the leaves were already on their way down, regrowth will be quick; if the leaves were still in good condition, you may have to wait awhile. But don't despair; unless the pot soil is overwatered you can be certain that the gloxinias will regenerate, for since that time I have not lost a single tuber from this treatment.

During the winter the plants exist in a sort of half-life, looking quite dreadful, with flabby, poorly colored small leaves. I keep gloxinias in a cold greenhouse where the light is good and the temperature stays around 45 degrees. Throughout this period it is extremely important not to overwater, for the roots are barely active. But don't go to the other extreme and allow the pot soil to dry out, for that may trigger off the unyielding dormancy I struggled against in the past. In bitter weather I take the precaution of using lukewarm water, and I am always careful not to get water on the leaves. In cold conditions, gloxinia leaves are apt to rot if they get wet, and this rot can spread to the tuber with fatal results.

From some crude tests I have since run, I now think I know what happened when I rested the plants in the conventional manner. Unnoticed by me, they sprouted even when they were in bone-dry soil. But when the tentative little growth was ignored it withered unobtrusively away, and after that the tuber would not resprout. The same thing will happen if you knock the new shoot off a gladiolus: although the corm seems unharmed, there will never be any more top growth. So if it is easier still to follow conventional practice for winter storage of gloxinias, watch them very carefully and repot them as soon as there is any sign of active life; with this treatment you will have no problems.

A greenhouse is not necessary for the plants when they are in this twilight sleep. A window that is cool and bright will do equally well; winter sun does not harm gloxinias. But the pots should be kept turned so that all the leaves get equal time with the sunlight. If this is too much trouble, make a reflector screen out of cardboard and aluminum foil and put that behind the pots. This often serves as well as all that turning. Indoors, be careful not to stimulate them with too much warmth or too much artificial light, for this may force them into faster growth than can be controlled successfully in a window and the plant will fail. Sometime in late spring you will notice that even without any stimulation there is a change in the appearance of the leaves: they will have stiffened up, turned a better color, and started to expand. At this stage any plant that has already flowered twice in the same pot soil should be repotted. Don't put this off; it is dreadfully hard to repot brittle-leafed gloxinias after the foliage has really got going. The growing medium is identical to that used for tuberous begonias, and again I add fish meal. It is equally important with gloxinias to make certain that there is plenty of internal drainage material in the bottom of the pots.

Plants that have blossomed only once in their pots can have the top two inches of soil scraped out and replaced with a fresh mixture; this top dressing will hold them for another season. But don't use fish meal in the top dressing for that will bring the feeding roots of the tuber up to the surface of the pot,where they are vulnerable to damage.

After repotting, my plants go back to the greenhouse but in a far shadier position, for spring sunlight soon gets too hot for the leaves. Indoors the repotted gloxinias should stand in a bright sunless window on a deep saucer filled with pebbles, for good humidity is now important. Just as soon as the danger of frost is over, the plants should go outdoors. In town I stand mine in a shaded cold frame with a glass cover propped open above them, so that drenching rain won't injure the house-tender foliage. By the time we move to the country in late May I have no further qualms about gloxinias; they go straight onto the high shelves in the lath house, where they have to take whatever the weather brings. The sooner gloxinias can be put outside the sturdier they will become and the more buds they will set. If by June a plant looks wretched, throw it out. The trouble is probably root rot, and there is no sense struggling with that problem. Watering should be quite regular all summer, for even though the plants may be outside, their huge leaves deflect water and the pot soil of a gloxinia that has just been through a torrential storm can still be dry. Never stand these plants on saucers outdoors; let them have a free runaway of surplus moisture. And be just as wary of overwatering as ever; pot soil can get sodden outside too. If the soil in the pot looks very dark and wet, make a lot of crisscross scratches to let in air and then don't water for several days. This often checks potential trouble. With such rich soil in the pots I don't feed my plants; indeed one of my minor problems is the enormous size of the foliage growth. If this begins to get top-heavy, prop it up with small stakes carefully inserted under the leaves and against the stems. But don't try tying a gloxinia up; the job is almost impossible to do attractively, and the ties often cut into the stems.

By August the buds will have formed, and these slowly lengthen their stalks to hold the flower above the leaves—a sign that blossom time is very near. In mid-August I bring the first pots with extended bud stalks onto the recessed display porch, where they are hurried on into bloom by the extra artificial light by day and the floodlights by night. By early September the full show is on, with every shelf loaded with blossoming plants. Again give plenty of water, particularly to any that may get some sunlight, but keep the water off the flowers, for it ruins them. If we stayed longer in the country I might try to stagger the flowering period by forcing even earlier blossom with artificial lights. But as things are, I let nature decide when the show will be at peak, and I get about a month and a half of concentrated bloom. If you have no covered place outdoors to show off your home-grown gloxinias, bring them indoors as the buds show color and put them by an open window. Once in flower, some protection must be given or the blossoms will be ruined by the rain.

After all this you may well wonder whether saving gloxinias is worth so much effort, and in some ways I suppose it is not. But I get enormous personal satisfaction from not wasting plants. To throw away something that can and should be saved because the saving may take time and effort is symptomatic of much of what I dislike about our modern outlook. And on the practical side you will never be able to buy any gloxinia that is half as large or a quarter as floriferous as the old ones you have saved year after year.

Wildlife

In the past few weeks a lot of my potted plants on the display porch in the greenhouse, and under the lath house, have been knocked down in the night, and in the morning I have found smashed tubers lying on the ground and earth tossed around everywhere. My first inclination was to suspect a panic-stricken bird, for the initial disaster took place in the small leaky shaded greenhouse where I raise seedlings and rest old plants. But now that the same thing has happened several times and in such different places, it is pretty clear that there is a definite plan to this attack, for it is the pots that contain the fish meal that seem to suffer the most, and birds, I am sure, can be ruled out.

Whatever animal is doing the damage obviously climbs well and is very neat, for there is nothing clumsy about the aftermath of these raids. Pots are not thrown down at random as though an animal had been prowling along the shelves, but some particular pot has been singled out for destruction. The problem in pinpointing the offender is the number of animals that inhabit this small lot of land—there are plenty of potential candidates for the black mark. It could be the family cat, who is a great night hunter and might possibly have forgotten himself and chased a frightened mouse or bird up onto my shelves. But I didn't think that he would have been so neat about the damage, for when his feral instincts surge through his veneer of civilized behavior, he is strong, rough, and fierce. But since he was making life impossible by trying to sleep on my bed the night of the last attack, he is now absolved.

It could possibly be skunks, of which we certainly have our full share. But I have never heard of skunks climbing quite so high, and I don't think that they could work in such confined areas without leaving some trace of their unmistakable smell, of which there is none around the smashed pots. We are all too well aware of the presence of skunks at night; occasionally I have had to shut my window because of them, for they are now hard at work digging holes in all the lawns in search of grubs—a useful action, though it does not improve the appearance of the grass; as I walk past the innumerable holes they have dug, I can only hope that I will be compensated by a huge diminution in the number of Japanese beetles in the yard next summer.

Raccoons are another possibility, for there are also plenty of them around the place, and these animals are not only great climbers but very neat-fingered. They also love fish, so the meal in the pots could have been a tremendous lure. Raccoons have increased considerably everywhere in recent years; they are not bothered by the proximity of man, and they have become quite a nuisance in some suburbs by raiding garbage pails. But I don't blame them for this particular problem, because their gigantic raids on my ripening corn suggests that they are finding all the special food they could possibly need, and when the corn is ready, raccoons prefer it to anything else.

Like everyone else who wants to enjoy his own garden-fresh corn, I have rather mixed feelings about this particular animal just at present, but I have recently read that a plastic bag slipped over the ear of corn just as it is almost ready or a sprinkle of red pepper on the ripening silk will deter raccoons, and if this works I will be able to enjoy these animals all year long. Last year a litter was raised in a hollow tree near the lath house, and very early on several occasions I saw the female, attended by all her little ones, hard at work among my corn stalks. But even as I mentally bewailed the corn, it was impossible not to be delighted that the tribe could find enough to support them on our land and need not venture into much less hospitable property alongside us. This year I have not found their headquarters, for they have abandoned the tree, probably for the same reason that we are going to have to take it down—it is so rotted that it is becoming dangerous. But I knew the raccoons were still around, partly because of the usual attacks on the corn and also because of the number of half-eaten grapes that I found thrown on the ground under the arbor. Then the other night, checking plants under the grape vine to see whether I should throw some plastic over the plants because there was a definite nip in the air, I noticed an enormous raccoon perched motionless on a high shelf beside some tuberous begonias whose pots were full of fish meal. Raccoons are related to bears, and although they are not wantonly aggressive they will attack if they feel cornered. This raccoon was large enough to remind me very forcefully of its family ancestry. So I beat a hasty retreat, leaving my plants to their fate and feeling pretty sure that I should find the tuberous begonias dug out of their pots in the morning. But in fact there was no damage; the raccoon clearly was after the grapes, not the pots, so I am now convinced that the raccoons too are in the clear.

I have also found to my surprise that we now seem to be supporting opossums, which I am again sure are not the culprits. I had always assumed that opossums were confined to the southern states, but the other night, hearing an unusual grunting noise, I opened the door and saw a pair of these animals in the courtyard guzzling down the fallen ripe fruit from the kousa dogwoods. My immediate reaction was to assume they were escapees from someone's private zoo, but I now hear that a considerable number have been sighted in this neighborhood in the past few seasons and the species is clearly on the march north. But although opossums will climb to sleep, I do not think that in this great migration northward they will have altered their normal habit of spending most of the time on the ground.

So I have now decided that the attackers are gray squirrels, which also haunt the garden. Gray squirrels pose quite a problem for animal lovers; they have so very little to recommend them. They do considerable damage to birds and to tree foliage, and heaven help you if they get into your attic. Nothing would ever induce me to try to eliminate any animal, even if it were possible, but gray squirrels are not favorites of mine, and this is yet another black mark. And yet we owe many of our trees to their activities, for it is from their buried hordes that nature provides us with new trees.

But there is no question that squirrels fit all the requirements for identifying the attackers. They climb, they are neat-fingered, and they are at present in their usual fall frenzy, trying to lay in winter provisions, which would make the old fish meal seem doubly attractive. In any case, the solution is an easy one: just as soon as I spot the squirrels working up that familiar head of steam next fall, I will move the pots that contain fish meal into a place that can be closed at night.

But in the course of this detective work I have been pleased to discover that our small woodlot and yard is supporting so much wildlife, for we also have chipmunks, occasional raids by groundhogs which ruin the vegetable plot, and deer that I occasionally see high-stepping their way across the wet grass in the early hours to get at the fallen apples from our old tree. Somehow we have all managed to coexist peacefully for many years, and the appearance of the possums shows that there is space for still more.

And no matter how much damage these other inhabitants of our gardens may do, I would not have it any other way. Today it is very difficult to give children the chance to see wildlife anywhere except in a zoo or on a reservation. We count ourselves lucky that there are still so many safe and surviving animals under our eyes in this small area.

Swan Song

In September, as the nights begin to cool off, the flower beds become their untidiest—and also their most colorful. The plants seem to sense from the falling temperature that their end is near and put on a final tremendous burst of bloom in a last extravagant attempt to attract insects to set seed and accomplish their ultimate mission in life, which is the continuation of the species.

Growing annuals in this area where there is not a very long season of hot weather is a lot of work. Ever since I have gardened in this country I have struggled with setting out dozens of little pre-raised seedlings in late May, cosseting them through the shock of transplanting, and bringing them to reasonably tidy bloom with stakes and string—not to mention warding off as far as possible the diseases and pests that lie in wait in every yard. Each year I resolve not to go through all this effort again, for spring regularly is marked for me by a sore back, broken fingernails, and roughened hands as the aftermath of the struggle. But the garden would be a poor place without flowering annuals; when they glow with color all those protestations are forgotten, and possibly I enjoy the final fling all the more because once again I did it!

It used to be a dreadful wrench when we had to shut the house and abandon the garden in order to get back for school. I would return to a dull flowerless yard and think longingly of all the flowers that I had left behind. One year I recall a local friend stopping in with an armful of my own annuals. She explained that she had passed the empty house and the plants were making such an extravagant, unappreciated show that she felt honor bound to pick some and bring them up to me. I remember that I received the bounty of my own labors with rather mixed feelings!

Nowadays we stay here until the frost puts a final end to the show, and early September is the time when I again become active in the flower growing area, where, truth to tell, I don't do much work during the summer heat. Annuals bloom reasonably well from July onward, and deadheading—that is, cutting off dead flowers as a routine process—is the proper way to handle these plants. I try, but I don't always get around to doing enough of this cleanup work until the urgent blaze of early September color warns me that the final burst is on. I then risk the bees and hornets, which are equally enthusiastically at work during this period, and energetically cut off every spent flower head that I can find, for with this treatment, dahlias, cosmos, salvias, scabiosa, petunias, marigolds, and nasturtiums—to name only the plants most usually grown—will continue to produce sheaves of flowers that will be larger and brighter than any that appeared earlier in the summer.

But I don't confine this deadheading to the annual plants; I also do the same job among the perennials, a necessity that is not stressed enough to beginning gardeners. Seed pods are cut off annual plants to encourage them to produce more flowers. Perennials need the same treatment, not to force more flowers, but to prevent the plants from exhausting themselves ripening seed, which is a debilitating process. But don't get so carried away with the clippers that you decide to cut off the rather battered perennial foliage as well. Day lilies, true lilies, heleniums, Michaelmas daisies, rudbeckias, shasta daisies, echinops, and phlox may all look a little wretched once they have finished flowering for the year, and you may feel that you are doing the plant a favor, as well as improving the appearance of the bed, by cutting off all the foliage, but this is mistaken reasoning. Perennials need their green tops, no matter how tatty these may look, for they are still carrying on the vital process of photosynthesis, which nourishes the plant against the coming winter. Cutting off perennial foliage while it is still green will not only deprive the plant of much-needed nutrients, but it may also send the poor thing into reverse. For if the weather stays warm and reasonably moist, a well-rooted perennial will not accept the removal of all the foliage without trying to make a comeback. Often it will send up fresh young shoots as though it were spring. When the coming cold weather slaughters these tender shoots, the old plants either die or, if they are exceptionally tough like day lilies, have to compensate for the loss of this extra foliage with far fewer flowers the following season. So no matter how wretched it may look, leave perennial foliage alone until it is completely dead. Then it can be cut down to four-inch stubs, for it is never wise to shave off a perennial plant right to the ground.

A little compensatory interest, such as a few of those budded hardy chrysanthemums that are now available on every roadside stand, goes a long way in the garden at this season. These are not indoor plants but field-grown chrysanthemums that can take outdoor living. In theory, therefore, they are hardy and carefully replanted should stand the winter outdoors and reappear next year. But this theory won't work for the pots you now plant outdoors. Had the chrysanthemums been set out in your yard when they were tiny and grown on in their flowering position all summer, then they would have come safely through the winter. But as it is, their roots will have no chance to take hold and the hard weather will kill them. So enjoy them for what they are and don't expect to see them next spring!

But no matter how faithfully you may have deadheaded your flowering plants all summer, you are bound to find a few nearly ripe seed pods as you work among them. And if your garden has done particularly well with very lavish bloom, the question inevitably arises whether it is worth saving some home-grown seed.

For most amateur gardeners the answer is no; your own seed will germinate poorly and prove a disappointment in color and form next year. If we could be transported back to the gardens of our grandfathers we might well be surprised by the poor color and quality of many of the plants we still use today. This is because we use seeds, or little plants grown from seed, that have been worked over for innumerable flowering generations by the plant hybridizers to produce better plants by controlled breeding.

The excellent seeds that are in use today are not for the most part the product of nature; they are nature-modified by man to give us greater pleasure. One example familiar to many gardeners is the annual Shirley poppy. We can sow a package of that seed and be certain that all the flowers from it will be in delightful pastel shades. But this form never existed in nature; the strain was produced through careful hand pollination by a clergyman who lived in the village of Shirley in England. Today every Shirley poppy is the descendant of a natural sport or mutation, a freak flower with a white edge around the normally all-red petals of the wild European poppy: an oddity that the Reverend W. Wilks happened to notice growing in a wheat field.

To produce those pastel-colored poppies took many years, for each successive batch of seed had to be culled of plain red flowers; only the flowers that contained some white were cross-pollinated by hand. Gradually the old red forebear was bred out of the strain, and Mr. Wilks did his work so well that the Shirley poppy seed is stable. None of those plain red ancestors ever reappear. But much of the present hybrid seed has been produced far faster (Mr. Wilks took decades with his work), and is therefore not so stable. In consequence home-saved seed, even if it is hand-pollinated from another flower on the same plant, sometimes throws back to a forebear that is much less attractive. And how many of us go around our gardens hand-pollinating our plants no matter how dedicated we may be as horticulturists? We leave this work to nature. Insects, bees, and butterflies are not interested in preserving the purity of a plant strain; all they are after is nectar. They don't stay confined to a single specimen, but move from plant to plant and from one color to another, which adds a multiple jeopardy to the mixture of genetic strains the seed will inherit.

I had a highly personal involvement with this particular problem with my sunflowers. At one time I used to buy expensive imported sunflower seed that was dwarf and interestingly rayed in dark red colors. No one else grew sunflowers in the immediate neighborhood, and the natural cross-pollination of the bees kept the strain pure because there were no variants around. But in recent years there has been a great increase in winter bird feeding, and one of the chief components of winter seed is sunflowers. Not every seed taken by hungry birds gets eaten; some are dropped and some carried quite a distance and then lost. And with this proliferation of feeders, a number of rogue sunflowers of all shapes and sizes began to appear all over the neighborhood.

One summer a few years back, one of these strays sprang up in my garden near an old fountain. And since it was the pride and joy of a visiting grandchild—and also because it was a considerable distance from my own special plants—I let it be, particularly as it was a dwarf variety and rather pleasant to look at with plain pale primrose flowers. Somehow its virile pollen got carried to my sunflowers, for the following year the plants in my garden showed the effect of cross-pollination by a marked change in their appearance. Though both parents were dwarf, the cross-bred plants did not inherit this characteristic; instead they grew much taller than in the past. The flowers were smaller, though still rayed, and the pale color had vanished; they also shed far more pollen when brought indoors, a sign that they were far more fertile.

I was sufficiently disenchanted with their appearance to decide that this was the end of the line and that I would buy new imported seed the following year. But the rejuvenated stock was not to be denied; although I tried to cut off all the seed pods an absolute thicket of self-sown sunflowers appeared the following spring. We happened to need a temporary screen along the back of the garden, and I used some of the volunteers for that purpose. It was a great mistake. Like the offspring of the household fly that is all the stronger because its parents survived spraying, the third generation of cross-pollinated sunflowers proved tough take-over artists that got out of control. They reached for the sky and spread so wide that nothing could be planted within yards of them. The flowers, tiny and plain and not worth cutting, were perched on enormously tall branches about twelve feet from the ground, and these branches were so brittle that they broke with the slightest breeze, littering the landscape like fallen giants. Rather in the manner of science fiction stories, my hitherto docile sunflowers suddenly turned back into one of their more deplorable ancestors. That was the end of them: I did not allow a fourth generation to get started. Indeed, I almost tremble at the thought of the plants that might have appeared!

Another familiar example of the unfortunate aspects of home-ripened seed is very commonly seen with the flowering phlox that is such a feature of the summer garden. Most of us have either heard of, or have had the experience of buying, some of the many excellent new hybrids of this plant, only to have it apparently revert to a peculiarly dull purple after a few years. But in this case it is a misconception to use the term "revert," for the expensive hybrid perennial has not changed. It has died out, having been crowded to death by its own seedlings. This happens when hybrid phlox is allowed to ripen seed, which then falls close to the mother plant. This seed is not stabilized, and the sturdiest little plants that come from it, which are the only ones that will survive in such unpromising conditions, are almost always a throwback to a basic purple ancestor from whose bloodlines the better plants have been bred. It is their descendants that we see blooming away in the flower garden and surviving in neglected yards.

For all these same reasons I also feel dubious about the advice I sometimes read that now's the time to search around among the blossoming annuals for little self-sown plants that can be potted up to flower indoors. To begin with, most hot-weather annuals detest indoor living, for there is not enough bright light for bloom. And if you are a good enough gardener to bring annual plants to flower on a windowsill, why not make certain of getting an exact replica of the plant you admire by taking a cutting of it, for that is the only way you can be certain of perpetuating an improved variety. But don't forget that an increasing number of hybrid plants are being patented, which means that the hybridizer who introduced them has the rights over the distribution of any cutting taken from his patented plant. Any such cutting can therefore be used only by you and never given away or sold. I do use self-sown seedlings from two of my own plants, the hanging lobelias and the *Primula malacoides.* The primulas are grouped together still in their pots outdoors in late April, and the fading lobelias are hung under the laths near other potted plants in early June. By late July, I have hundreds of self-sown seedlings in the worn-out soil of the potted primulas, and I can find as many lobelia seedlings as I need growing happily away in the wet gravel on which the house plants stand. From these I pot up the smallest specimens, following the old rule that with hybrid plants the best results come from the little seedlings, and in this way I have grown generation after generation without buying fresh seed.

But I have success with these two plants because I grow each species in only one color. And no one near me grows any of the same plants, so there is no cross-pollination by insects from nearby inferior stock of another color or with a different characteristic of growth. Nevertheless, I get an amazing variety in the size and color shadings of the flowers, for even these interbred seeds that have not had any new blood injected into them for more than ten years still carry different genetic qualities inherited from their most distant ancestors.

Just as long as you realize the problems ahead of you, experimenting with home-saved seed can bring about some interesting results. What it will not bring about is a garden of uniform quality, which I happen to want in my flower beds. For that reason I clip steadily from now until frost, and loudly as I may bewail the cost, I buy new hybrid seed each spring.

Country Gardens

One of the pleasures of being a gardener comes from the enjoyment you get looking at other people's yards. I learn a great deal when I visit fine display gardens, although other people's immaculate lawns lined with brilliant flowers often give me severe twinges of pure envy. But I get even more pleasure watching the month-by-month changes in the little yards that front the roads that I drive. For I watch these gardens at every stage, at their best and at their worst, and although the owners are strangers, the portions of their gardens that I can see from the car are familiar friends. Gardeners are a generous lot: they willingly share their knowledge as well as their cuttings, and an interest in plants transcends any personal involvement. I therefore get enormous pleasure from the sight of a new yard being constructed and feel considerably depressed when I see a piece of land falling into neglect. These reactions are, I am sure, common to all horticulturists.

Among the gardens that I count as old friends is a very simple plot that consists of only a lawn with two flower beds planted with petunias lining both sides of an unusually long driveway. My back ached in sympathy this past wet spring when I saw the householder out on his knees in the pouring rain setting out dozens of little plants. I was on my way to do exactly the same job in my own yard, for the season was running so late that I dared not wait any longer to get the annuals in. In the past, this particular gardener had always used petunias in mixed colors dotted in among each other for his display. They grew well and were carefully kept up, but the overall effect was always spoiled for me by the muddling up of the colors. This year, to my delight, there has been a change of taste, and the petunias were set out in blocks of solid colors. This was such an improvement that I dropped an unsigned note of congratulations into the mailbox, for who of us does not like to have our experiments applauded?

Farther along the same road there is another house that has a round single flower bed by the front door. This has always been planted with hot-weather annuals that don't give much of a show until early September; I feel sure that they bloom so late because the flowers are grown from seed shaken over the ground, for the plants have that unmistakable sturdy look that comes from never having been transplanted. This season the bed is now ablaze with zinnias in every possible shape, height, variety, and color, which makes me fairly certain that home-saved seed must have been used, for many examples of ancient ancestors that are part of the genetic makeup of modern zinnias are flowering gaily in that bed alongside their descendants. Possibly all this diversity of form makes for interesting flower arrangements indoors, but outside the jumble is distracting, which is a shame with such well-grown plants. I can't wait for that gardener to have a change of heart and stop frugally using home-set seed. Maybe I could speed up a new horticultural deal for the little round bed by mailing a package of modern hybrid zinnias to that house!

Nearby there is an absolutely plain, quiet, green yard that is well kept up but makes no effort about color except at this season. And then suddenly, year after year, a double row of bright red salvias comes into bloom beside a long fieldstone wall. The dramatic use of a single color in plants of uniform size and shape against the soft gray of the wall is always breathtaking, and each year I wait with considerable anticipation to see the show light up. Once it comes, it lasts a long time, and this burst of glory in the usually quiet yard is all the more attractive for being so unexpected.

There is also a vegetable plot beside that road which I have enjoyed watching. This, I am sure, is a family plot largely worked by someone who first grew plants in Europe, for the land is laid out very much like the small plots that can be seen in Portugal or Italy. There are no flowers in it at all, which would certainly exist if the gardener was of British extraction. The area, which is not large, lies on poor soil beside what was until very recently a heavily traveled road. By good husbandry the soil has been continuously enriched and, judging from the crops, kept enormously fertile. Fumes from the cars were minimized by a tall wire fence over which cucumbers and climbing beans were trained. Everything was laid out in neat rows and the earth between them was regularly hand-hoed. Growing there were all the usual staples as well as a large asparagus bed, a big patch of strawberries, raspberry canes, blueberry bushes, and off to one side a grape arbor.

I never saw anyone working in that yard; I think the owner must have labored over it at night when he came back from work or extremely early in the morning. It must have provided enough vegetables to feed a large family with plenty left over for canning, and there was never any roadside stand beside it to suggest that surplus produce was sold—everything pointed to family use only. But, alas, I have to use the past tense about the perfection of that little plot, because this summer I noticed to my dismay that half of it is no longer cultivated. I would like to think that the family had grown up and there was no longer the need to grow so much. But gardens are very revealing, and this one shouts out the news that there has been some disaster in the family.

Someone is trying to keep that garden going on a much smaller scale but without the skill that was so obvious before. The asparagus bed has vanished, the raspberries and grape vines are wildly unpruned, and the disused half is a frightful mess of weeds, which would never have happened if the plot had been deliberately reduced in size. I would almost prefer to see the whole area abandoned and returned to the natural meadow from which it once was carved than watch the pathetic attempt that is still under way to make something of that once beautifully managed plot.

But while my old friend is, I am afraid, slowly on its way out, a new vegetable growing area has appeared quite close by, alongside a cranberry bog. First a small new house was built, and then instead of the conventional lawn the owners put in a huge stand of corn. This amused me, and I applauded such independent thinking.

But the following season I noticed the owner out rototilling more land, and a wonderful vegetable area has been developed as well as a huge strawberry plot, plenty of raspberries, and corn. But the corn is now planted more conventionally to the side of the house, and there is a small orchard with dwarf apple trees in front of the house. This garden is not enclosed in the conventional sense, but its limits are defined with lines of floribunda roses which produce a very sprightly effect. It shows the same concern for the land and the same desire to get the best out of it without mistreatment that I admired so much in the old-fashioned plot, and since I am apparently losing my old friend, I am delighted that I can have a chance to watch the development of its modern equivalent.

When I realized that the old vegetable strip was on its way out, I suddenly recalled a very distinctive flowering plot that I used to admire along a back road we took to town before the modern highway was built. That road ran through deep country, and many of the houses were totally unimproved. At one turn in the road a piece of land that could hardly be dignified by the name of garden was filled higgledy-piggledy with some of the most spectacular annuals I have ever seen. This strip was near a pond, and its owner was a very elderly lady whose derelict little house with the shutters hanging loose was still lighted by lamps. I don't think there was any running water in it either, for I used to see the old lady pulling up water in buckets from the pond and taking some indoors and using the rest on her flowers. There must have been something very special about that pond water, for even though the flowers were scattered around with no rhyme or reason, and were also mainly the products of home-saved seed, the flowers were bigger and brighter than any that I have ever grown. Thinking about it made me go to look for it, but, alas, the plot is gone. The road has been widened, the broken-down cottage swept away, and that fine talented natural gardener is no longer there. I shall always be glad that I saw that extraordinary array of flowers and that I frequently stopped to discuss them with the old lady. For she was lonely and her flowers were her children and she loved to have them noticed. That, she told me, was why she grew them beside the road.

With the widening of the road, some building has begun and a little way past the vanished garden I noticed a small brand-new house with a yard that was obviously a great source of pride. It had a newly made lawn, young trees had been set out where they would comfort the house with afternoon shade when they grew taller, and dozens of white petunias bloomed in every open space. The effect was extremely lively, with a youthful look, and a baby carriage on the breezeway confirmed this impression. I was entranced by the spontaneity of the place and stopped for a better look, and it was then that I got a nasty shock. Like most new houses, the front was dominated by a large picture window, and there on a pedestal in proud glory was a frightful arrangement of plastic leaves and flowers.

The incongruous combination of the well-planted yard and those flaccid, fraudulent flowers continues to haunt me. I don't want to think that love of living plants and love of plastic flowers represent differing tastes of husband and wife, for that would suggest the possibility of even more dangerous incompatibilities in matters more important than flowers. To comfort myself I have come up with an alternative explanation. That frightful bouquet must have been given the young couple by some important relative, and for diplomatic purposes must be on display for a time. But I do hope that the statutory period during which it has to be exhibited is nearly over, for as long as it gleams so prominently in the window, a delightful yard is being spoiled.

Fall Duties

One of the pleasant, leisurely jobs I enjoy in the fall is cutting down the deciduous ground-cover plants, particularly the hostas, variegated goutweed, and day lilies that line the open areas around our brick courtyard. This is easy work, requiring no mental effort, and I keep an old-fashioned sickle on the porch so that I can chop down a few feet whenever the spirit moves me. But every so often I have to give up the steady swinging swish of the sickle and get down on my knees to pull out hundreds of intrusive dogwood seedlings. These come from the nearby kousa dogwoods and are rapidly seeding themselves not only in the open areas but also between the bricks of the court itself. Considering the excellence (and cost) of kousas as decorative trees, and the speed at which they turn into flowering specimens, it seems a shame that

I can't hold a garage sale of these little trees or even invite people to come in and dig their own as blueberry growers do on their farms. I save some each year for myself, and I have a large collection. The oldest, which have been in place for seven years, flowered for the first time this season. But this entire area would not be large enough to contain all the volunteer kousas I find each year, and lovely as a glade of them would be, I haven't the space to grow so many. They are deep-rooted and take some getting out without heaving the bricks if they are left more than a single season, and in their profusion, they will soon rate with violets as a lovely but pestilential nuisance!

And no matter how you may procrastinate, this also is the time to clean up the flower beds. It is admittedly a traumatic experience to destroy plants that are still in bloom—pulling them out seems a poor recompense for the pleasure they have given. But it is better to steel yourself and get the job done before frost does it for you. For later, as it gets colder, the bed will look frightful, and because of the weather you will be even less willing to do anything about it.

After the annuals have been consigned to the compost pile and the dead perennial stalks cut off, you will feel as though a great burden has been lifted from your shoulders, but don't put away the wheelbarrow, for the end is not quite in sight. You will do the garden, and yourself, an even greater favor if you will follow the cleanup with a very thorough weeding of all ground nuisances, the withered remnants of the crab grass, purslane, ground ivy—whatever is your land's particular menace.

Far too many gardeners feel that the fall work on the flower bed is finished when it is cut down, and leave the ground to go into the winter full of weeds. This is the direct route, horticulturally speaking, to a miserable spring, for many of the worst garden weeds are not annuals that will be killed off by the cold, but perennial plants whose roots will be active next spring, far earlier than the gardener can get to work, turning themselves into bigger, more troublesome pests that will be twice as hard to get out.

Working outdoors pulling weeds in the spring is a very different proceeding from puttering around doing the same job in the balmy days of early fall. In the spring the soil is cold and wet, and trampling over it in that condition can do its structure serious harm. Where I am concerned, it can also do me serious harm, for while in the fall I am in fine shape and can undertake a hard day's weeding without turning a hair, in the early spring I am soft and prone to the miseries if I work long outdoors!

And don't feel that this weeding is hardly worthwhile because of all the weed seed that you know is lurking in the soil ready for action next year. If you cover the clean ground with a four-inch-thick layer of mulch—either leaves and grass clippings which you can get for free in your own yard or wood chips and fir bark that can be bought at garden stores—the beds will look pleasantly neat during the dreary months ahead, and the mulch will prevent regrowth of the weeds if there is a prolonged Indian summer. Since some of every mulch breaks down or decays over the winter, the layer will not be thick enough to prevent the regeneration of every spring-sprouting weed seed, but it will make a very great difference in the number that germinate and also make them very obvious for easy elimination before they get thoroughly established.

Fall is not the end of the gardening year; it is the start of next year's growing season. The mulch you lay down will protect your perennial plants during the winter and feed the soil as it decays, while the cleaned-up flower bed will give you a huge head start on either planting seeds or setting out small plants.

For gardens, unlike the rest of us, do not have to accept the inevitability of old age. If we will but help them a little, each spring will find them just as young as ever.

October

Regeneration

The little piece of rough ground that leads down to our waterfront has provided us with a fascinating grandstand view of the way land in this area regenerates after a natural disaster. The area slopes slowly down to the harbor, and at the lower end there is a brackish salt marsh into which a small brook flows and over which the tides slide daily.

Before the 1954 hurricane the marshland was full of reeds, bayberry bushes, and cattails and loud with peepers in the spring. Where the land starts to rise toward the road, but was still vulnerable to occasional abnormal tides, there were crab apples, tamarack, tupelo, junipers, and white ash, all of which seem able to survive an occasional dose of salt water at their roots. Above the tides there were oaks, birches, swamp maples, white pines, and a few enormous old red cedars.

The hurricane ruined it all. The wind knocked down every tree, the trees in turn smashed down the bushes underneath them, and the tidal wave that followed buried the marsh under mud and the litter of broken trees and shrubs under gigantic drifts of seaweed. At that time the land was not ours, and since the owner was old and ill the wreckage was never cleared out but remained a wasteland of decaying seaweed and noisome debris scoured from the ocean floor that smelled to high heaven.

It took several years for even the rankest weed to find a toehold in this unpleasant area. But slowly, as the stream and tidal waters cleansed the marsh, life returned to that piece of open ground, its renaissance aided greatly by foraging flocks of half-wild Canadian geese which at that time occupied part of the waterfront area. Geese are excellent weeders. Their big bills open up the ground, and although that flock fouled the open waters because there were so many of them, their endless scavenging activities undoubtedly speeded up the regeneration of the salt marsh, for it recovered from this fearful desolation far faster with their help than it did some years later when by mistake I nearly killed it a second time after the geese had departed.

But the geese did not forage in the upper area near the road, and the smothering drifts of seaweed acted for several years like a gigantic mulch, preventing any plant from starting into growth. Those that eventually got in were the invasive vines, and once they had reestablished themselves among the debris, the area soon turned from a wasteland into an impenetrable jungle. A wild profusion of Japanese honeysuckle, grapes, poison ivy, and bittersweet bound all the fallen trees together like helpless hostages. Nothing and no one could get through that thicket except the birds, and the area was full of them. The litter of branches provided excellent cover and nesting areas, while the vicious thorns of the smilax that had also reestablished itself kept both dogs and cats at bay.

At this stage we acquired it, and since it was a fire hazard and also blocked our route to the water, we decided to clean it up even at the expense of the birds. The tangle was more than any amateur could handle, and professional tree men took more than a month to clear out the wreckage of all the ruined trees. But by the end of the growing year we were left with a large new vacant lot, and I began to make plans for replanting.

I need not have bothered with the plants, for the moment the sunlight again reached the ground all kinds of new life appeared. The first were huge clumps of daffodils and May apples—both of which must somehow have survived years of frustration under the litter but which reappeared bigger and better than I would have thought possible. Tiny little cedars the size of matchsticks popped up, and innumerable sprouting acorns proved again the tenacity of life in long-buried seeds. But all this hopeful new growth did not have much of a chance, for by midsummer the strangling vines had reappeared in force, apparently rejuvenated (as so often) by the savage cutting out. But this time we did not allow them to get the upper hand. The area became the special province of my husband, and he watched over it with the greatest care, quartering the ground regularly to pull out vines and marking every new worthwhile little bush or tree with a stake. And they needed to be marked, for although the vines could be pulled out as they appeared, nothing could stop the regrowth of the worst possible rank grass that filled in the cleared area at lightning speed and had to be hacked at with a hand sickle wherever it threatened to overwhelm adventurous new trees.

This attack and counterattack continued for several years, and while the vines and the rough grass were so strongly on the offensive nothing else did very well, for it soon became evident that if the vines were to be eliminated their roots would have to be got out. And grubbing up the roots invariably dislodged some small tree that had been carefully staked for safety but which unbeknownst to us was straddling a dangerous underground root.

In the end virtue triumphed, we got the vines under control, and the shrubs and trees were able to make some growth. As soon as these were large enough to shade the ground, our problems with the grass vanished, for it was a variety that needed full sun to flourish. By the time the junipers and oaks were beginning to reach a height that was at least noticeable, a second stage had set in. We began to find clumps of white birches, crab apples, privet, and philadelphus, all of which had grown in the area originally but which we feared had vanished completely. Some ground elder which had survived nearby also moved in and has been a pest ever since.

It is now twenty years since that particular disaster, and after a lot of vigilance on my husband's part that area has completely regenerated. The first of those little junipers is over five feet tall, the crab apples have started to bloom, and on the waterfront a number of rugosa roses have sprung up, presumably from haws washed ashore in high tides. The better trees have also made a strong comeback; there are quite large tupelos, plenty of red and white oaks, horse chestnuts, some seedling hollies, and even some white pines that did nothing for many years but are now growing very fast. The vines are down but not out; every season a huge tangle has to be sent away to the dump, but with the thickening up of the cover, the nesting birds have returned.

But it has been a very long, slow process that without our active, vigorous intervention might never have taken place, for it would have been years before those fallen trees decayed sufficiently to allow the buried seeds sufficient light to sprout. And had they done so, they would never have turned into anything, for they would all have been strangled by the vines.

This experience showed that while destruction can take place in a matter of hours, natural regeneration takes decades and may indeed never succeed without artificial aid.

The potential of land to redeem itself continues for many years under even the most difficult conditions. The weeds that spring up in the broken cement of old cellars bear constant witness to this fact. But spoiled land must have help to bring it back into a natural combination of growth, and the longer it is left without this help, the less chance there is that good trees and shrubs will ever reestablish themselves. Land overwhelmed with trash vines and litter sometimes cannot regenerate except with deliberate replanting. For in nature, as in all things, good does not necessarily triumph. Therefore, for any of us who own land there is one absolute necessity—that we keep constant watch over it.

Chrysanthemums

The plant windows are full of flourishing pots of home-grown chrysanthemums and a fine long-lasting show is now starting that will continue until December, for every plant is loaded with buds that will open. Sometimes if the hard frosts have been delayed and I have been able to keep the pots outdoors late I can keep the display in good enough condition to have it still on view as part of the Christmas show. When this happens I am following a tradition of my childhood, for I grew up before poinsettias, cyclamens, and poor overforced little azaleas were used as Christmas plants. The sight of tall chrysanthemums full of flowers being carried into the conservatory was always the signal that other Christmas delights were just around the corner.

I never bring or use potted chrysanthemums indoors before October, although I cut the late-flowering garden varieties for the vases from September onward. Potted chrysanthemums in flower are now available commercially throughout the year; last spring to my dismay I even saw some combined with red geraniums, coleus, ageratum, and a few pingling white petunias for Memorial Day baskets. Chrysanthemums have become flowers for all seasons and for every occasion since the professional growers learned to manipulate the bud set by changing the light hours.

I am not very happy with this special forcing: no matter how skillfully plants are grown in the forcing houses, the process is an artificial one sustained only by very elaborate equipment. Forced out of season, a plant may produce flowers, but the buds that have also been set are not going to come into bloom when the pot moves into your house, and the plant will loathe the change in growing conditions. This is as true for a chrysanthemum as it is for a gloxinia. If you want to get your money's worth out of a plant, buy it at its natural flowering season for longer life.

But there's more to the matter of out-of-season chrysanthemums than just a short life span. Chrysanthemums are etched into my subconscious as a symbol of fall. I don't want spring chrysanthemums any more than I want daffodils in November, for with all due respect to the skill of the commercial growers, I prefer my flowers at their normal blossoming season. This is a personal reaction and unimportant to the flower-buying public in general, but I do wonder why, since chrysanthemums have now become an all-around staple of the florist trade, they are offered to us only in such dull varieties and such uninteresting colors? Part of that question I can, I suspect, answer. We are offered these uninteresting chrysanthemums because these are the types that force the best under unnatural conditions. But that does not seem to me a good enough reason to explain away the fact that we are not offered much better varieties at the natural period of bloom for these flowers.

Better types are around, as anyone can see for himself by looking in the refrigerator cases in a florist shop, but they are offered only as cut flowers and not as potted plants. Better plants might cost a little more, but if the dealers would but try them out they would probably sell like hot cakes, for people like something new. Some of the more interesting varieties of these plants are slightly harder to grow—they do better in greenhouses than indoors—but I have never noticed that this fact bothers the retailers where other plants unsuited to house living are involved. And as for the difficulties of managing spoon, spider, incurved, or mop-head chrysanthemums indoors, could they be any more trouble-some (not to mention impossible) than gardenias? I think that we should at least be given the option of trying out some of the vastly improved pot chrysanthemums that have come on the market in recent years. If we are warned and we fail, we bear no grudge. Indeed, in matters horticultural I have made plenty of errors over the years, but in the process I have learned a great deal so I cannot see why we should be spoon-fed by the retailers in this matter of chrysanthemums and offered the most boring varieties of a very exciting plant, just because it is the easiest for them to raise—and ostensibly the best for us to look after indoors.

A great many people hardly know that more interesting shapes and colors even exist as potted plants of this species. This is brought home to me every year by the comments I receive about my own chrysanthemums, which to people conditioned to the dreary varieties available in the shops seem highly unusual. But I don't grow any that are particularly out of the ordinary; I merely grow more interesting specimens than those on sale, and I have never found these the least hard to raise or to keep going for months on end indoors. The process is easy and can be carried out successfully without a greenhouse by anyone who owns a small plot of sunny ground. All that is needed is to buy cuttings in late spring either through the catalogs or at a garden center. I choose late-flowering varieties and usually the kinds listed as dwarf, for chrysanthemums always outgrow their suggested height in our rich pot soil, and can turn into stalky plants. I confine my choices to two varieties in, at the most, three colors—one of which is always either white or cream-colored. I don't buy mixed collections of cuttings, for although these often include very interesting plants, the different types flower at slightly different periods and come in a great variation of size. I aim for a uniform effect.

Basically the process consists of planting the cuttings in small pots if it is not yet outdoor growing weather and keeping them in the brightest coolest place possible. When they have recovered from the repotting, the growing tip should be nipped out; nail scissors do an excellent job. As soon as the weather warms up in May, they can be knocked out of their pots, which by then are full of nice strong white roots, and lined up two feet apart in any area that has full sun. If the soil is poor, scratch in a good peppering of dry cow manure before the plants are set out. Once they are safely planted out, the soil should be covered along the rows with a thick mulch of half-decayed leaves and grass clippings. If you have no leaves, grass clippings alone will do equally well, but keep them away from the tender stalks for a day or two after they are laid down, lest the heat generated by damp spring grass injures the stems. The mulch will keep the ground cool and moist even through prolonged dry spells and also prevents weed growth, and as it decays it provides additional nourishment for the roots. The growing points should be pinched back: once at planting-out time if there has been good branching after that first potting-up pinch, and again if there are at least three more inches of new growth before the really hot weather sets in. The books suggest no pinching after July 4. I pay no attention to this, for I want my plants late and full of buds, so I continue to pinch until I have a well-branched plant.

After that nothing more is done for my plants except keeping the mulch renewed by emptying the grass catcher from the lawn mower near where they grow and spreading the fresh-cut grass around them. If there is a prolonged drought, and I notice the plants wilting badly, I may take the trouble to drag the hose over and give them a slow soak. But when there is a deep mulch, wilting is hardly ever serious; it takes place only during the heat of the day, and the plants recover during the night. Watering is more likely to do damage in a humid heat wave than a little wilting. For if the foliage of chrysanthemums goes into the night damp, gray mildew may appear.

After Labor Day the plants will have become large and bushy

and should be lifted on a cloudy day. The most painless method is to take a barrowload of soil and all the necessary pots and shards out to the growing area. Crock the pots; I use six-inch plastic azalea pans and put in a bottom pad of rich soil dusted with dry cow or steer manure. Water the growing chrysanthemums lightly and then dig them, driving a spade in a circle straight down beside each plant about five inches away from the main stem. Don't work underneath the root ball until the circle is complete. For the actual lifting I use a small shovel, for the longer handle gives me more leverage. Inevitably, some earth will fall off the root balls no matter how careful you may be, and sometimes the roots will hardly fit into the prepared pots and have to be rammed in, doing some injury. The repotting process should leave an inch between the soil surface and the top of the pot, and no extra soil should be added except around the sides where there may be some gaps. Thump the pots up and down several times on the hard surface to make sure that the roots are in contact with the new soil in the base of the pot.

Put the newly potted-up plants into deep shade, water the soil, and spray the foliage with a weak dilution of some water-soluble fertilizer. Don't rewater the pot soil until it looks a little dry, but mist the foliage lightly each morning. Most chrysanthemums move without turning a hair if you are reasonably careful, but occasionally they wilt alarmingly. In that case don't panic and pour water over them; keep up exactly the same treatment you are giving the sturdy plants, and in time even the most wretched misery will recover. The move invariably kills some of the lower leaves, and cramming the plants into pots crowds the leaves in the center so that they get no light, so some will die there too. Dying and dead leaves should be regularly pinched off, for as they are removed, more light reaches the interior, which gives the survivors a better chance. Immediately after the repotting tie up the foliage, which may be flopping all over the place. For this use thin bamboo green stakes, three to a pot, inserted almost invisibly and slanting outward in a triangular shape from the center of the plant. Green wool is best for the actual tying in and will be almost unnoticeable.

About a week after the initial repotting the plants should be tested in a brighter place. If they wilt badly, take them back into the shade. Eventually they will all be able to handle full sun all day, which is the way they should continue to be grown. Put saucers under the pots at this stage, for the roots, suddenly confined, are ravenous for moisture to keep the foliage in good condition. Often water is needed morning and evening. Once a week a light feeding of water-soluble fertilizer is also needed.

Leave them outdoors sheltered from strong winds as long as possible. The light first frosts will not harm them if you throw plastic garment bags over the pots after the sun goes down. Pots that remain outdoors late set more buds and are sturdier. But bring them indoors as soon as the buds begin to swell or show color and before hard killing frost.

These garden-raised repotted cuttings will make superb indoor plants in a bright cool place. They will flower for weeks on end, and every bud will open. But when eventually the show is over, throw the plants away. They can be saved, but it's a long tiresome business even with a greenhouse. You will do better to start with fresh cuttings in the spring.

But what about people who can't raise their own plants: is there any hope for them? At present not much, I am afraid, where unusual varieties are concerned unless a special order is put in with a garden center in the spring. Not every garden center will accept such a commission, but it's always worth asking. Whatever you buy, choose a plant that is only just beginning to show color —don't fall for one in full bloom or you will not get much for your money—and keep the pot in the brightest coolest place you possess on a deeply pebbled saucer, going easy on the watering. I wish

I could offer better advice to the gardenless gardener about this plant, but the choice remains disgracefully limited. Nevertheless the solution lies in our hands. The only way to get better plants is to demand them, and if enough of us do this regularly, while spurning those available, the growers will eventually get the message.

I appreciate all the work that the experts undertake to give us better and stronger flowers, and I am quite ready to salute them for their skills. All I am asking is that they give us a chance to buy some of the results of all their work on this particular plant.

The Extravaganza

It so happened that until recently I had not been in England for many years at the period when the leaves were starting to turn. But last year when I was there at that particular season I was struck afresh, after the long absence, by the delicate manner in which that landscape prepares for winter in contrast to the flamboyant last fling of a New England fall.

In England, autumn comes on quietly and almost imperceptibly as a soft mellow period in muted tones. And as the leaves start to turn, they shine out on misty mornings in tones of rusty browns and soft yellows against a pale sky. Very little red is to be found among the leaves of the native trees. Red does exist in the autumnal coloring, but mainly in the form of berries in the hedges where, deep among the thick hedgerows along every back lane there are bright clusters of red hips and haws, poisonous nightshades, and the scarlet hues of unripe blackberries. But where our berried bushes are protected for the exclusive use of the birds by menacing tangles of poison ivy, a fireworks display in itself, the berries of the English hedges are guarded by somber ranks of stinging nettles in olive drab that diminish the impact of the color. So here the general effect of fall is stimulating, almost invigorating, while autumn in England, as many poets have reminded us, is a quiet period of gentle melancholy.

It must have been an extraordinary experience to see the colors of a New England fall for the first time with no advance publicity—and a great pleasure to be able to write about it in terms that were not yet clichés—and yet the early settlers made very little of it in print. Initially, of course, they arrived too late in the season to see the show, and by the time the color display appeared again they had been through a great deal. But even so, those fiery colors must have come as a revelation to them, and in spite of their silence on the matter, I can't but hope that some of them found time to rest from their incessant struggles to provide for the colony and enjoy the display which surely must have cheered their anxious spirits.

The early settlers turned their backs deliberately and determinedly upon their homeland; they sailed away from it with few regrets. But homesickness is a curious thing; it has a habit of creeping up silently and unexpectedly and overwhelming you like a wave when it is triggered off by sight, smell, or sound. I wonder whether the older folks may not have suffered that sudden unmistakable surge of nostalgia when suddenly they saw the trees flame up like a glittering candle against a sapphire blue sky and recalled the quieter colors they had left behind forever.

It all must have come as such a shock. At first the winding down of the growing season must have seemed very familiar, a gradual blending of brown and yellow as the ferns withered, the goldenrod faded, and the cranberry vines turned russet brown. The ruby colors of the leaves of the Virginia creeper may have been the first hint that something more was on the way, but no one could possibly have expected the high drama of the fall foliage, for nothing that anyone had seen previously would have prepared them for it.

In spite of the distance and the time taken by the sailing

ships, there was plenty of traffic and correspondence between the Old World and the New, and the settlers wrote freely about the natural wonders of this strange land. But although they wrote of another of our fall phenomena, the hurricane, and commented extensively on the return of the migratory birds in the spring, nothing has yet been discovered that suggests the slightest excitement about the extraordinary impact of the changing leaves. Perhaps they thought the colors hard and barbarous, a reflection of the difficulties surrounding them in their new life. Or could it be that they distrusted all foliage color after an initial contact with poison ivy?

Whatever the reason for their silence, I shall never believe that a New England fall could have gone unnoticed, for it has always been part of the heritage of this area, something we love and praise. But there is a question that we soon must face concerning our continued enjoyment of this natural extravaganza—do we cherish it enough to make sure that this same loveliness will be available for our descendants to enjoy too? Pollution, roads, lumbering, the unending increase in paper products, building, and the overuse of winter salt against icy roads are all taking a heavy toll on our trees, particularly the sugar maples, the keystone of the color spectrum. People today drive for miles to see the foliage when the color is at its height. Are we prepared to involve ourselves enough in conservation matters to make sure that the same beauty will be there for people to admire a hundred years from now?

South African Bulbs

Plants, like all living things, have strong preferences—not to mention positive requirements—about what best suits their particular life style, and careful growers make great efforts to meet these needs to get good results from difficult but worthwhile plants. But, as I have learned to my cost, even the most exact attention to highly picky needs is never an automatic guarantee of success, and that hard fact is extremely well illustrated by the off-and-on performance of my nerines.

Nerines are South African bulbs related to the amaryllis family and often inaccurately described by the all-embracing name of Guernsey lilies. This is one of those happenstance names that gets misattached to a plant in the first place and then becomes the generic term for the entire species—just as we speak of Kleenex, meaning tissues, whether we actually buy the Kleenex brand or not. The plant that got the romantic name of Guernsey lily (which by the way has nothing whatsoever to do with Lily Langtry, who was called the Jersey Lily after a different island among the Channel island group) is technically *Nerine sarniensis.* It has crinkled flowers that are such a bright pink that they are almost red.

The colloquial name got attached because a shipment of these bulbs washed ashore on Guernsey from a wreck some time in the seventeenth century. In the island's mild climate they naturalized themselves to perfection in the sand above the high-tide mark. No one, by the way, ever thought they were native to that area; it was apparently realized from the first how they had established themselves. But since the ship from which they had been lost had come from Japan, the plants were supposed to be of Oriental origin. Not until the late eighteenth century were wild specimens located growing in remote areas of South Africa, so it is now assumed that the boat picked up a shipment on its way home.

In New England the entire family has to be grown under glass, and they all have bothered me for years with their highly temperamental behavior. For nerines will not adjust to life in another hemisphere. They insist on doing things the way they are done at home in South Africa, which means that the bulbs rest dry in their pots in the summer, bloom in our fall, which is their ancestral spring, and then send up a clump of straplike leaves that have to be kept growing in good health all winter. And they will do none of this indoors unless the plant window is as bright and as cool as a small greenhouse.

All this would be fine and bearable for those who own greenhouses and like to grow striking and unusual plants, if only nerines would do what is expected of them. But unfortunately they can be exceptionally troublesome to induce to bloom, and more than one excellent amateur gardener of my acquaintance has thrown in the sponge in frustrated disgust and stopped trying to grow any of the family. I have still not quite reached that stage, although I have been very near it several times, only to repent when, after sulking themselves to the verge of the trash can, my pots suddenly reward me with a long-lasting show of attractive pink wands with multiple flowers at the tops of the stalks at a season when there is very little else other than chrysanthemums in bloom.

I also cannot quite bring myself to get rid of some pots that, combining all the worst features of their usual unpredictability, have made an out-of-character adjustment to living on this side of the world. These are some misguided nerines that a few years back refused to stop growing in the spring and stubbornly continued to throw up flourishing green leaves. Normally a plant that does not drop off to rest naturally can be forced into dormancy by slowly withholding water, but not these oddities. They went on producing new growth even when the soil was bone dry, and eventually I gave in and started to rewater. In early October, just as their relatives were coming to life with a few buds at the neck of the bulb, the misfits started to go dormant, but as the leaves yellowed off they too budded and flowered in a hideous tangle of dead leaves, and they have continued this curious pattern of flowering at the usual season but reversing their resting period ever since.

Nerines are available in the British catalogs in the most enchanting ice cream colors that I long to possess. Indeed, if the flowers on a home-grown plant are allowed to go to seed, these develop into flowering-size bulbs very fast, so it is perfectly possible to do some home hybridizing among the various types if you can induce your bulbs to flower. But on the whole, it is a mistake to be too ambitious with this bulb; don't try to run before you can walk with it. I have not yet tried any of the expensive imported bulbs, though I have flowered some of my seedlings, for I want to be quite sure that I have mastered the basic demands of this finicky bulb—and that certainly is not yet at hand!

If you want to try your luck (and you may well succeed where the experts fail), buy the bulbs at this season and pot them in deep, well-crocked, but small pots. Remember these are members of the amaryllis family and they should therefore not be overpotted. An inch between the bulb and the edge of the pot is more than sufficient. Leave half the bulb above soil level and use rich pot soil. Nerines increase extremely fast by means of offsets, and since they loathe root disturbance it is better to provide them with good nourishment from the first so that the evil day of dividing can be put off as long as possible. The reason why nerines are such problem dividers is because the roots, like lilies, never go completely dormant and are extremely fragile. It is better, therefore, to move an entire clump up two pot sizes if it has been flowering well rather than pull it apart. If you insist on dividing the bulbs, do the job while the plants are in their resting period but very near to the time when regrowth will begin. This is a tricky thing to estimate, for the roots start to whiten and swell in the dry pot long before there is any sign of top growth. Always knock a clump of nerines out of their pot and look carefully at the roots before dividing—if any roots look active it is already too late. It is also a good idea never to divide all your potfuls at the same time. Stagger the work over several years; otherwise you may have no bloom at all, for it can take nerines as long as three years to recover from division.

Old pots should be started back to life by being sunk under water in a bucket and left there until the air bubbles stop rising. There is no other way to make absolutely certain that a bone-dry clay pot is completely saturated. If it is not, the dry clay will steal much-needed moisture from the pot soil. Often flower buds will appear without any watering, but in this case the tub treatment is even more necessary, for the bulb has exhausted its reserve of moisture and is throwing a bud without any nourishment being available from the soil.

Buds normally appear on a long bare stem with no leaves visible. But there are varieties that send up leaves and a bud simultaneously. And with most of us there are innumerable pots that send up leaves without a bud! When the flower fades, cut it off but leave the stem to wither away naturally. Seedmaking exhausts the bulb, and it is more sensible to wait to try this until there are numerous flowering pots in your possession. All through the winter the green leaves must be kept growing in a very bright but cool place with regular watering. Some form of water-soluble fertilizer should be added every second week. By May, unless you have offbeat plants like mine, the leaves will start to fade, and the pots should be rested in a warm place for at least four months. I leave some pots upright in the cellar where the temperature never goes above the 65-degree mark, and some I rest on their sides on a sunny shelf in the greenhouse where the temperature is often above 100 degrees. I have never noticed any difference in the subsequent behavior from either set of bulbs. The oddities I rest in the winter in a warm room near the furnace alongside the amaryllis. If no buds reappear after all this effort (your new bulbs, having been specially prepared, will certainly bloom), you will have joined the large majority of those who grow this exasperating plant! But when and if flowers do appear, they are such a triumph that all is forgiven—and that, I suppose, is why I am at this season peering daily into the shriveled necks of my bulbs to see what kind of a flowering year it is going to be this time!

But not all South African bulbs present these problems, and one, the lachenalia or Cape cowslip, is delightfully easy. Lachenalias are far more tolerant of human frailty than nerines, and they can even be grown in a sunny window indoors if—and it is a big if—there is an unobtrusive spot where their rather battered foliage can continue to grow all winter, for they, like nerines, must be allowed to carry their leaves for a long time after bloom if they are to produce flowers the next year.

The bulbs, which usually have to be bought through catalogs, appear on the market in September, and they can safely wait for several weeks in a cool dry place before they are potted up. Lachenalia bulbs are small and can be crowded into a plastic bulb pan, for they are not deep rooters, and they can be grown in commercially packaged soil given a better texture with damp peat moss and perlite. Set the bulbs, which unlike nerines will have no root system to worry about, an inch below soil level and put the pots immediately into a sunny cool place. There is nothing to be gained by starting growth off in the dark. The larger *Lachenalia superba* has a rocket-red bloom and rather large leaves, and this variety should be unobtrusively staked. *L. superba* flowers in the very cool greenhouse three months after it is planted. *L. tricolor,* a smaller bulb that throws out little spotted leaves that do not need staking, carries drooping yellow flowers in March. This plant looks much better seen from below and does well planted in one of the new plastic hanging pots that have a saucer attached.

After flowering, the foliage must be kept growing until it withers, an uninteresting period but one which will bring far better blooms the following year if water-soluble fertilizer is given the pots every other week while the leaves are in good condition.

When the foliage is browning off, the bulbs can rest in their pots. But unlike nerines, the resting place should be reasonably cool; lachenalias wither away if they are rested in a hot place.

Ours summer bone dry in the cold cellar, and when we return to town in mid-October there are usually shoots already above ground in the worn-out pot soil. These shoots, like those on the gladiolus, appear before any root growth starts, so it is simple to turn them out and make up fresh plantings. In so doing you will find that there are dozens of little offsets clinging alongside both *L. superba* and *L. tricolor*. These minute bulbs usually will flower the first year as a miniature replica of the old bulbs. There are two methods of handling them, and I use both. Some bulbs I repot with all the infants still clinging to them. These produce a nice multiple clump of flowers the next year. Others I break up and use to make small pots of new bulbs which I force-feed with weekly doses of liquid fertilizer. Both methods are always entirely successful, and you will end up by giving bulbs away by the handful.

Lachenalias are not the most spectacular or the sweetest-smelling of the bulbs I grow, but they are by far the least trouble-some and well worth trying for yourself.

Don't Stop Yet

Fall in all its mellowed fruitfulness always seems a lovely leisurely period. The children are back in school, the days are warm and sunny, and we can work outdoors in our yards or in-doors puttering around the plant stand without any sense of pres-sure, tuning our relaxed mood to the winding down of the grow-ing year.

But all this serenity is a deception practiced upon us by nature, for now is in fact one of the busiest seasons in the gardening calendar: there are many jobs to be done, and some that can be done *only* at this time.

If your lawn is weak, this is an excellent period to clear out weeds, for early frosts will have killed off the crab grass. Any cleaned-up bare lawn areas can be topped with sterile loam bought at a garden center and new seed put down. Grass seeded now may not produce much of a show before cold weather sets in, but the seeds are in fact sending down deep new roots even though there may be nothing above ground. There will be a stronger, better stand of grass in the spring from fall-sown seed than from a spring seeding.

This also is the one and only time to set out the spring-flowering bulbs. For people with small yards, the little bulbs—snowdrops, scillas, crocus, chionodoxa—are the best investment. If they are planted just below ground where there is some sunlight but not hot, baked, dry soil in the summer, the little bulbs will reward you by flowering year after year and increasing hugely in numbers. Little bulbs are a bore to plant, but if you take a shovel and open a large shallow planting area, a great many can be set out in a big clump that will be all the more effective for the massed effect it will produce at flowering time.

Daffodils and hyacinths should also be planted now. These too need a place with moderate sun where there is reasonable moisture, but the soil must be neither swampy nor packed hard and dry—and where the dying foliage will not infuriate you when the flowers have gone. Again I prefer these in clumps, and in small gardens I don't like the huge old-fashioned varieties such as King Alfred. There are early-flowering daffodils, particularly a variety called February Gold, that look much better in a limited area and have the additional advantage of coming into flower sooner than most daffodils and lasting longer. These look exceptionally well planted with plenty of blue scillas around them, and although I do not usually favor mixing various types of plants together, golden daffodils and blue scillas make excellent exceptions.

Tulips should also be bought now while good varieties are still available. If you live in a windy area, try a planting of the single or double early tulips. These do not grow nearly as tall as the later tulips and are not nearly as vulnerable to bad weather. They are also far the best tulip for the beginner to try to force indoors. Tulips can be had in varieties that produce an extended season. But the late Darwins, everyone's mental image of the perfect tulip, and the late lily-flowered tulips, lovely as they are, bloom at a time when there is often violent wind and hot sun, a combination that either snaps off the flower heads or bleaches them out. All new bulbs come with the flower bud already safely established inside; there is no way that the bulb can be prevented from flowering the first year, so if you insist upon growing Darwin or any other late tulips, take advantage of this built-in bloom which does not call for a particularly suitable growing position, and plant the bulbs in a sheltered place in light shade. This is heresy to serious growers, for it means that these particular bulbs will not ripen their foliage and set bud another year. But since tulips are a chancy business the second year at best, I think it more sensible to plant new bulbs where they are safe from wind and dehydrating sun and be sure that you will have a good show with no thought for tomorrow.

If possible, hold off planting your tulips when first you buy them; early October is too soon for them to go into the ground. Store them for at least another three weeks in a cool place, with the bags torn open to admit air. Tulips start to grow too soon if they are planted while there is any chance of a prolonged Indian summer, and if premature top growth is subsequently exposed to bitter winter cold the tulips may be killed.

I usually finish off the outdoor planting season by setting out the tulips under miserably uncomfortable working conditions.

Each year I vow I will not go through the frozen-finger routine again, but each spring I am glad I did, for tulips flower more evenly and reliably with late planting.

For indoor gardeners hyacinths are the simplest to grow of the Dutch bulbs, but only if there is a cool, dark area available where the pots can be stored while roots are made. In recent years I have taken to rooting my potted hyacinths and daffodils in the back of the cold cellar in plastic garment bags, which I huff and puff up so that they look like small zeppelins. The buds root as well there as they do in the cold frames outdoors under leaves, which was my old routine. The big advantage of the cold frame is that bulbs can be held and taken out slowly for a staggered display. Rooted indoors, the bulb pans are inclined all to come into clamorous growth simultaneously. This drawback has to be set against the problem of getting at the cold frames at all under enormous piles of frozen snow, and there is no question which method is easier on the back! But for controlled forcing—for flower shows in late spring, for example—I would always use cold frames, for while growth can be hurried on with extra light and moderate warmth, bulbs that have peaked too early cannot be held back.

If you are a gardenless gardener living in a high-rise apartment where all this talk of cold frames and cool cellars belongs to a never-never land, use paper-white narcissus or Soleil d'or or French Roman hyacinths for bulb bloom. These can be planted in succession for steady winter bloom in plastic bulb pans, and they will do better for you if you use ordinary commercially packaged soil. These bulbs do not need any preliminary dark period for root forcing but can go straight into a sunny window. But do stake the foliage as soon as it elongates; bulbs grown indoors are usually burdened with lanky leaves and should be tied in carefully with green wool. It is important to lay in the winter supply of paper-whites now, when they are available, for they vanish later in the winter. Store them in mesh onion bags in a cool room—or if planting is delayed, put them in the crisper of the refrigerator.

House plant growers should also take this season as the moment for being exceptionally strong-minded about their plants. Anything that is shabby in the fall will look deplorable by Christmas and impossible after the New Year! Throw it out and buy something new, for winter indoors is never a recuperative period for any plant. This, by the way, is not the right season for repotting house plants; leave them alone and compensate for crowded roots with an occasional mild dose of liquid fertilizer. Repotting should be put off until spring creeps over the windowsill.

For holiday flowers, the Christmas cactus should be rested now. This is not the same plant as the Thanksgiving cactus, and the difference shows up in the tip ends of the modified stems that serve as leaves. In the Thanksgiving cactus these stems have sharp pointed claws, and this plant will soon set bud without any special effort on your part. But if the ends of the stems are rounded and blunt, the plant is a Christmas cactus, which can be troublesome to make set bud. A Christmas cactus will never bloom if it has to spend the night in a temperature over 70 degrees, so if you keep the thermostat high, resign yourself to using the plant for a foliage decoration. In a cooler house, this species can be induced to bud if it is taken to a dimly lighted place for the entire month of October and left without water and without any light at all—neither a streetlight nor even your coming into the room to find a book—after six o'clock in the evening. If your plant is so tightly packed into the pot that there is no soil left, a slight misting of the stems during this enforced rest may be necessary. But for most plants this should be a completely waterless period. By the end of the resting period, which should never be less than four weeks and sometimes can take as long as six, there should be tiny buds showing on every blunt stem end.

Horticulturally, therefore, there is plenty to do right now. Happily, there is very little stress involved with it at this delightful time of year. But do get going—cold weather could set in unexpectedly early—for what you do now about your indoor and outdoor plants will set the tone for the success or failure of your personal horticultural season next year.

November

Conservation at Home

The fact that leaf burning is now forbidden in many states seems to be upsetting people quite unnecessarily. The fuss shows how far many of us are from understanding the problems of conserving the fertility of the soil, let alone the interrelationship of man with the world in which he lives. For since the ban went into effect in our area, not only have I seen more and more advertisements for garden attachments that will "grind up or pulverize all your leaves so that they can be easily disposed of in plastic bags," but I have also been made aware that certain trees are now considered less desirable than others because of their very heavy leaf drop each fall.

Impeccable neatness in a yard, without a leaf in sight and with all the natural debris of nature shredded and sent out, does not necessarily indicate good husbandry. On the contrary, it is evidence that the owners have no conception of the natural process by which land regenerates itself year after year without the need of expensive fertilizers. Or if they do have an inkling, they are not prepared to act upon it for fear of neighborhood pressure about what constitutes a neat yard.

I totally agree that fallen leaves in some areas of a garden look dreadful and should be cleared away. Driveways and paths should not be left covered with leaves, since this makes them slippery and dangerous. Lawns left covered with leaves are also a sad sight—the grass beneath will go into the winter with a severe disadvantage. Fallen leaves pack down and make a smothering mulch that can kill grass, and most of us have enough trouble getting something approximating a lawn in the first place without willfully facilitating its all coming to nothing by failing to rake up the leaves!

On these matters we are all in agreement. The real difference lies between those who feel compelled to get rid of all the leaves in their yards and are disturbed because they can no longer be disposed of with a wasteful match or sent out in unlimited quantities to our overloaded dumps and those who want to help in the area of conservation, are aware that leaves play an important part in the recycling process of nature, but still don't quite know what to do about the piles in their own yards.

Leaves need not be a nuisance to even the most tidy-minded gardener. Indeed, rather than being bothered by their abundance, we should be thankful that we live in neighborhoods where such problems exist, for leaves are friends at every season. There are ways of turning them to good use after they fall, and while they were alive they functioned as the route through which the vital life process of the bush or tree was sustained. Most of us realize that this planet could not be inhabited by man, beast, bird, or fish if no oxygen were available in the atmosphere—some of our space probes have been designed to discover whether any of the nearby planets are equally fortunate. But not all of us are equally aware that oxygen is not a free gift that just happens to exist here; it is manufactured for us by the green surfaces of trees, plants, grass, lichen, mosses, and algae, and if these were to be completely destroyed all over the earth no other form of life could then survive.

Our dependence on living plants is what conservation at its most fundamental level is all about, and the interrelationship among everything that lives on this planet is never better illustrated than by a full understanding of what leaves mean to us—and we to leaves.

Plants give off oxygen as a waste by-product of the process in which they convert moisture and carbon dioxide into carbohydrates—starch and sugar—necessary for their growth. They achieve the energy to carry out this process through a still-imperfectly understood system that is triggered by light falling upon the green substance known as chlorophyll which exists in leaves and stems. While the minute openings on leaves and stems are passing out oxygen, they are simultaneously taking in carbon dioxide from the surrounding air. Since this particular gas is a serious offender in the ever increasing problem of air pollution, the presence of trees, shrubs, and grass in any neighborhood means that the air will be purer around them. I am well aware that this is an oversimplification of an extremely complex process, but it is all most of us need to know, since the effect of green things on any area is very obvious.

If, for instance, you have ever walked from a street lined with trees or shrubs onto the asphalt of a parking lot, you know how much hotter and more lifeless the air immediately feels. This partly comes from the hard surface underfoot and the absence of shade. But trees not only protect us from the sun; they also give off tons of water daily through their leaves, thus lowering the temperature in their immediate vicinity. And by walking out from under the trees into a barren area with nothing green growing in it, we are also walking into a much greater concentration of polluted air, which adds to our sense of discomfort. So, if there were no other reason for appreciating leaves, their power to filter the air should make us overlook the temporary nuisance they turn into when they fall. But even this should hardly be counted as a nuisance, for dead leaves are invaluable in your yard if you will but use them instead of looking around in wild despair to see how best to get rid of them.

As a start don't clean up the place so thoroughly. Let the leaves remain where they drop under shrubs and bushes as long as they are not obtrusively untidy; concentrate on raking them off only essential areas. By the following spring, if you are a doubting Thomas, you will discover that those you left have decomposed into a thin rich mulch that is adding immeasurably to the fertility of your land.

Once you have proved to your own satisfaction that leaving the leaves under the bushes is not going to deteriorate the value of your property, you are ready for the next stage, which is to pile extra leaves into the shrubbery and around your specimen bushes each fall. The amount of leaves that go down is a personal judgment; I don't like them piled on the ground so deeply that they still retain their original form the following spring. What you needed to discover is the depth of leaves that will get broken down in the course of an average winter in your particular area, and never let them rise above that level.

If you have no shrubs that would do well with this natural mulch, find some out-of-the way position in your yard where all the surplus leaves can be piled up into a heap. If you are afraid they will blow about, an inexpensive wire cage can be made to contain them. But most yards have some odd corner where the leaves can be stored without presenting any problems. The only place where they should never be heaped is against wooden walls, where the process of decay will also attack the wood. Piling leaves is the simplest form of compostmaking; it is completely odorless and needs no effort on your part other than hauling the leaves to the designated spot—nature will take over from there. And nature works extremely fast, for you will find by spring that the overwhelming pile of leaves has shrunk tremendously, so you can go on adding to it almost indefinitely. But not without any return, for there is a bonus available from this simple process in the form of dark, rich leaf mold that can be dug out of the bottom of the pile year after year. And there is nothing better for your lawn, potted plants, or growing areas than homemade compost.

Leaves are nature's equivalent of "all this and heaven too" —and we should be properly grateful, for without them life would be far less pleasant.

Memories

When you live for many years in one place it becomes full of memories and associations that can prove a serious stumbling block to innovation. With the end of the outdoor gardening season at hand, I have recently been busying myself with plans for updating the suburban garden to bring it more into line with my present taste—and also the amount of time and energy I now have to work in it. But at every turn my schemes are being thwarted by the close ties that exist between certain plants and the various stages of our life here, and I am not sure that I can steel myself into deliberately obliterating so many personal memories.

There is, for example, a rather battered bush of bridal wreath in the backyard that is rapidly turning into a disgrace, for it insists upon spreading so widely each season that it has to be annually hacked back to let light into the area nearby. This savage pruning is an exhausting chore which must be done at exactly the right moment, which is as soon as the flowers fall, otherwise there will be no blossoms the next year. This particular moment happens to be a fearfully busy time, for we are just getting ready to move to the country, and the rough treatment given that bush because of the press of so many other things to do has left it jagged and misshapen. If my new ideas are to come into being it must come out, and the area around it go down to paving—indeed, considering its appearance, the bush should come out whether my new plans come into being or not.

But close to it, and therefore destined to be eliminated if the huge root ball is extracted, are some of the very first bulbs I ever planted in that yard. They went in during the worst period of the German blitz against London, and every year when they reappear I am reminded of the stability and bravery of my kin when to the outside world everything seemed hopeless and lost. I am not at all sure that I want to forgo this reminder.

The new plan also calls for the removal of a tired evergreen hedge that runs beside the path to the back door. I planted most of this on a fearfully hot fall day while anxiously awaiting the diagnosis on a quarantined and hospitalized sick child. It was during one of the serious polio epidemics—a terror that modern parents no longer have to live through—and I finished the job after a telephone call that told us that our worst fears were groundless. I never look at those bushes without reliving the overwhelming feeling of happiness that followed that news, and how can I deliberately erase such a memory?

Beside the garage stands an apple tree that is an outstanding example of inappropriate, badly designed planting. The branches have constantly to be cut back to prevent them from breaking down the power lines that provide electricity to the greenhouse. And the ever increasing shade it casts has inhibited bud opening for the past five years on some rather nice hybrid rhododendrons set out underneath it. But our house and yard were originally laid out in an old farm orchard, and when we first moved in, a few ancient trees still survived. One of these I deliberately had cut down—it was so riddled with carpenter ants that we feared it might be dangerous, and its position made it impossible to plant a vegetable garden, a high priority since the war had started. But before the power saws got to work, I promised myself and that old tree that there would always be apple trees in the yard. I planted the present nuisance as its substitute. Foolish as it may appear, I don't feel like breaking that promise.

And where I have made changes they have not always been approved. The big front lawn where the children played baseball and which was tented over for all their weddings is redesigned, and the place where they danced is now a dwarf shrub border. After I had made these alterations I discovered that our daughters rather resented the elimination of what was to them a familiar part of the landscape of their childhood.

It is far easier in many ways to design a new garden than to redesign a familiar family gathering ground. Change in the suburban garden when it comes will have to be slow and unobtrusive.

Amaryllis

I am, of all things, being harassed by amaryllis (*Hippeastrum*) bulbs, probably because these are not great favorites of mine. In spite of their spectacular bloom, I have never cared much for this particular bulb. Their enormous size is a problem in the small pots their culture demands, the flowers do not last very long nor, to my mind, combine well with other potted plants, for they are so overpowering in themselves, and the ungainly strap-shaped leaves need long aftercare when the flowers are finished, unless you mean to throw an expensive purchase away. So after once proving to myself that I knew how to manage these South Africans I have admired them in other people's collections and left it at that.

But now retribution has set in, owing no doubt to my take-it-or-leave-it attitude. The story began two years ago when a young neighbor was given an amaryllis for Christmas. She is a busy person with young children and not much spare time. When the flower which I had noticed adorning her kitchen windowsill faded, I saw the pot put out on the back stoop on a bitterly cold day. It was clearly destined for the trash can, for it was extremely battered, with a leaf held together with Scotch tape. Since it had given me vicarious pleasure from my kitchen window I felt sorry for it and suggested that it be given to me instead of going out with the rubbish.

Last year it repaid me by flowering extremely well, though in an unfortunate shade of scarlet, and so, most mysteriously, did another totally unknown amaryllis that somehow, unnoticed by me, had insinuated itself into the greenhouse. Where that second plant came from I have no idea, and I am still wondering whom I may have offended by not writing my thanks! Then a visiting daughter who lives in Pakistan decided to buy a white amaryllis to take back to her garden. Amaryllis bulbs are much grown in that country, but the stock is poor and the colors muddy; a pure white specimen would be a star. Unfortunately her purchase refused to remain dormant and threw up two fat buds before she had packed her suitcases, so it too passed into my care.

The final straw came last week when a friend found unexpectedly that she had to spend the whole winter in Florida and gave me, out of the kindness of her heart, four of her best amaryllises as a thank-you present for boarding some of her other plants. Nothing could have been kinder, but a population explosion of amaryllises is the last thing the greenhouse or the plant window can take, and I too shall have to look around for friends who would appreciate a plump-budded amaryllis.

But for those who do not share my views, these are excellent bulbs to give as an expensive gift to a gardenless gardener who does not have a cool place where the usual spring bulbs can make those essential roots. What's more, a newly purchased amaryllis is a foolproof bulb, for the flower bud is already safely tucked away. To get a spectacular show, all that is necessary is to plant it in the center of a small pot with less than three-quarters of an inch between the bulb and the pot sides, leaving more than half the bulb protruding above the soil. Watering should be moderate until some sign of life appears, which with a new purchase is usually the thin blade of the bud. Then more water should be given, although the soil, which can be any commercial mix you find at the florist, should never be allowed to become dark and sodden.

The plant should be grown in a sunny window and turned regularly so that the elongating stalk does not bend toward the light. Some growers stake their amaryllises, but there is not much leeway for a strong stake in such a small pot without piercing the bulb itself—with of course disastrous results.

If the bulb was brand new and enormous, don't throw it out after the first bud fades, even if you don't intend to try to carry it on. Cut off the finished stalk and continue watering for at least another month. Those huge, preconditioned bulbs usually throw up a second bud after there has been a little leaf growth, and this second flower is just as spectacular as the first. But after the second bud has flowered that is the end of the line, if you don't want to be bothered with aftercare.

If you do want to try your hand at carrying the bulb on, plant it in the same small-size pot, but make sure that there are two overlapping pieces of broken pot covering the drainage vent in a clay pot, and a curved piece of shard over each vent in a plastic pot. Since an amaryllis needs such a small pot to bloom successfully, it is not possible to use the usual layer of shards. But for long-term success good drainage is very important, and the drainage vents must be kept clear of soil that could block them up. In a long-term project the bulb should also be given better soil: commercially packaged material can still be used, but the texture should be coarsened with damp peat moss, and a heavy sprinkling of sand or perlite should be added for good drainage and to keep the soil particles open so that the roots can find plenty of oxygen.

A couple of inches of the improved soil mix should be spread over the shards, and then a half-inch layer of fish meal should be worked into it. Fish meal is a well-balanced fertilizer that suits almost every plant, but it is not as long-lasting or slow-acting, whichever expression you prefer, as raw bone meal, which some growers still recommend as their first choice for fertilizer. The bulb should again be planted so that at least two-thirds of it is above ground. The soil mix should be poured around the sides of the pot and firmed in with the fingers; don't ram it in—this can injure the bulb, and amaryllis does not take kindly to very hard potting.

Grow the plant exactly as before except that when the flowers fade cut them off but leave the stalk to wither naturally away. Its green stem surface sends down much-needed nourishment to the depleted bulb. Cutting it off for neatness' sake is, if you want flowers another year, exactly like cutting off your nose to spite your face.

Once the flowering has finished, your main aim should be to get as strong and flourishing a growth out of the leaves as possible. This is best done by keeping the pot moderately watered in a warm sunny spot while the weather is still cold, and keeping it in full sunshine, if necessary on the fire escape, all summer! Throughout this rather dull period an occasional dose of water-soluble fertilizer is very important. The leaves will continue dull but healthy until some time in August, when they will begin to yellow off; this is a sign that the plant is going dormant. Not much can happen to the leaves over this rather boring period except attacks by a red spider if the pot is kept in a very dry atmosphere and a mild danger from mealybugs if you have other infested plants. Mealybugs attack amaryllis by getting under the onionlike outer skin of the bulb itself, and if this happens great harm will be done. So if you know you are plagued with this pest on the plant stand, keep the amaryllis in another room.

As soon as the leaves are clearly on their way out, the amount of water that is given can be slowly decreased. But it is essential to time this decrease to the withering of the foliage and not initiate the collapse of the leaves by withholding water. An amaryllis that has been hustled into dormancy by premature drying out will not set a flower bud.

When all the top growth has shriveled, watering should cease entirely, and what by then will be a deplorable-looking pot with bone-dry soil should be rested. I keep mine in the room where the furnace is located, for this is one of the few bulbs that must have a warm dormant period.

Most amaryllises need a rest of at least four months without any water from the time the foliage dies down. Unlike other South African bulbs, amaryllises don't usually start top growth without encouragement in the shape of rewatering, although in every collection there are always a few oddballs! It is better not to start rewatering until after the New Year, and it is a good idea to turn the bulbs carefully out of their pots beforehand to see whether there is any sign of root action. An amaryllis that is through dormancy will have swollen white roots; if nothing of that sort shows, the bulb is not fully rested.

Amaryllis dislikes root disturbance, so if you feel compelled to repot, do so before there are new roots to be disturbed. Remember that the soil mixture should be exceedingly dry; don't slap wet soil against a dormant bulb, or rot may set in. Most home-saved amaryllises are not nearly as large in later years as they were at first, so there is usually plenty of space in the original pot. If the first potting soil was very rich, you will get better results by shaking out loose soil alongside the bulb and pouring in a new rich mixture at rewatering time rather than attempting the radical operation of total repotting. As with nerines, it is better to stagger repotting of a collection so that some bulbs are undisturbed.

It takes about three weeks for the top growth of amaryllis bulbs to respond to rewatering, and not all the bulbs come back at the same time even if they went into dormancy simultaneously. But as long as the bulb itself remains firm to the touch, and not squishy or powdery, be patient; top growth eventually will appear. If you have beginner's luck and get a flower bud, don't be discouraged if it is a lot smaller than the original bloom. The flowers will never again be as big as those specially conditioned for the market. But if they seem outstandingly puny, water-soluble fertilizer should be given more often during the aftercare period. If there are no flowers, the problem could be premature dormancy, too cold a storage area, or too short a resting period.

No matter how you look at it, carrying on an amaryllis is a long-drawn-out job, but if you have room, and patience, it is worth trying to get more out of such an expensive bulb. The fact that seven flourishing potfuls are more than I can handle is surely no reason for anyone else to condemn a worthwhile plant!

Individuality

Plants can be extremely exasperating. In theory they should follow the fixed immutable laws of nature and act in a predictable manner if certain fixed procedures are followed to the letter. In point of fact they do nothing of the sort, for plants are highly individualistic objects. Even cuttings from the same bush that are raised identically can show variations of growth patterns. So any gardener who feels that he knows exactly what to expect is undoubtedly a beginner; old hands have long since learned to bend with the wind, take things as they come.

Outdoors, of course, the vagaries of the weather can be blamed for some of the peculiarities that occur, but this is not really the whole answer, as the inconsistent behavior of potted plants can prove. Not long ago I needed to produce a good specimen of air layering, the process by which a tall single stemmed plant can be induced to make roots high on the stem and produce a new small plant instead of a ceiling-tapping giant. I had not taken any layers for many years, and since there was a lot of new information available about such work, I restudied the procedure very carefully, found a suitable healthy plant, and followed the instructions to the letter. I was also particularly careful to choose the appropriate season for the job. The result? That plant managed to embarrass me publicly on TV by having no roots at all on the stem when the great moment for unveiling it under the lights took place!

The subsequent year I needed a sequence of slides of the same process. I took an identical plant, but otherwise I did everything wrong. Officially it was much too late in the season, the plant was in poor health, and I gave it a hard time by not working fast but leaving the critical wound areas exposed at every stage to take photographs. Operation over, I slapped a spagnum pad over the cut, covered that with tightly tied plastic, and bundled the unhappy plant into the cellar while I took a holiday. And with everything crying out to be done on my return, I did not get around to looking at the plant for nearly six weeks after I had operated on it. When eventually I rescued it from the cellar, the pot soil was almost bone dry and the leaves looked a little sad, but the plant was still alive and, amazingly enough, there was a lusty crop of white roots visible through the plastic threading their way into every part of the spagnum pad. I was in fact able to take a final photograph that I had never expected to include. That particular young plant is now starting to look as though it too will soon need an air layer, and the decapitated stem has thrown up two new shoots, which in itself is something unexpected for a *Dracaena.*

Gardenias are also plants that are completely unpredictable, at least where I am concerned. One year on my birthday my daughters gave me a matched pair of magnificent gardenias full of bud and bloom that had clearly been greenhouse raised to their enormous healthy size. My indoor record with gardenias is outstandingly bad, for nowhere in the house appears to suit them, and I normally keep mine in the greenhouse. But it so happened that at that particular moment there was no room in the greenhouse for those two huge plants, so I took a chance and put them on display in the main plant window, where there is controlled humidity, which gardenias need, but where the temperature can fall very low in cold weather, which is not something gardenias like. But with their usual inscrutability those plants flourished in spite of the fact that my birthday is in January and the temperature in that window by night was in the forties. They succumbed to all the miseries that infest gardenias only when eventually room was found for them in the far more suitable greenhouse!

This behavior so disgusted me that the following spring I took some cuttings from a gardenia grown successfully for years indoors by a friend of mine. I reasoned that if her plant was so house-tolerant—and she keeps her rooms much hotter than I do—it was a variety with which I too might have success. Three cuttings took and grew away successfully all summer under the grape arbor. If you must grow gardenias as house plants, you will always find they do better for you in the summer than when the heat is on in the house.

I needed those little gardenias for a TV demonstration, so I kept a very sharp eye on them. I repotted them the moment they needed it in carefully balanced rich soil that contained a magnetized iron element to ward off chlorosis—a yellowing of the leaves, to which gardenias are particularly prone, which results when the root hairs do not get the necessary trace element of iron in the dissolved salts that are drawn into the plant. True to the extraordinary individuality of gardenias, the three pots, even at this early stage, developed rather differently. One grew well and formed, during late fall, a number of buds that remained green and as tightly furled as an Englishman's umbrella. One did not grow so fast, but it also formed buds which opened with sweetly scented flowers that lasted a long time. The third plant became an outstanding foliage specimen with no buds. All this was fine with me, because the point I wanted to make on air was the astonishing variability of gardenias, and I also wanted advice from the viewing audience on how they handled theirs.

After the pots had had their turn before the cameras, I took them back to the town greenhouse, where, since the heat was not yet on, living conditions were very similar to the life the gardenias had lived under the grape arbor. The return to town marks the start of a desperately rushed period in that garden. Bushels of bulbs need to be planted and tidal waves of leaves swept up, so although the greenhouse plants were regularly watered there was no particular intensive care, for all the plants that had gone into the greenhouse were in excellent condition.

But when the worst of the rush was over and I had time to look carefully at the gardenias with a view to carrying out some of the hints I had received, I found to my annoyance that everything had changed. The specimen that had refused to open its buds outdoors was doing beautifully inside. The leaves were in fine shape, and there was a pleasant glint of cream among the buds that were slowly opening. The plant that had flowered outdoors was unhappy; the buds that remained had dried up and the leaves were turning yellow with chlorosis. And that fine budless plant was covered with the triple demons of scale, aphids, and mealybug which did not exist on any other plant in the greenhouse! And this from three cuttings from the same mother plant that had been raised identically and had stood side by side outdoors and in the greenhouse!

This is what I mean by suggesting that every horticulturist should be prepared for anything to happen with plants. Good culture does not always bring good results, although naturally it helps make success more likely. There are variables that affect plant growth which we still do not entirely understand. If you take up gardening outdoors or on the windowsill you must be optimistic, flexible, and prepared to try anything. Don't stick to one process or one set of instructions. Shop around among the many informative books available until you find what best suits you and your plants and your gardening conditions.

Exasperation

We recently had a rather odd fall with good growing weather continuing well into the end of November. Among the many abnormalities produced by this rare warmth was the fact that the grass continued to need cutting—something I found almost impossible to induce those who do that chore for me in the suburban yard to believe, for they had long since beaten their mowing machines into snow plows. In consequence the lawns went into the winter far too long, and their appearance the subsequent spring was very depressing. Mowing does not have an automatic shut-off date on the calendar. It must be continued as long as the grass grows; otherwise, when the snow falls, long matted grass will be pressed down against the roots, depriving them of oxygen.

Our normal practice with the grass, apart from keeping it short, is to cut out the dead places where crab grass has taken over after the frost kills it and reseed those areas, and anywhere else that seems weak, with the best grass seed we can buy. Since we get back to town rather late, this job is not usually done until late October, but this does no harm. For although the air is too cold for new grass to appear above ground, the roots make good growth in the still-mellow soil, which gives them a head start on spring. But during that long-drawn-out fall, the newly seeded grass did sprout. Long before the snow arrived I could make it out as a green fuzz. This raised grave fears in my mind about the likelihood of such tender top growth being able to survive the coming winter, and if the new top growth was killed then the roots would perish too. And those fears proved all too correct, for when the snow melted no new grass appeared.

Worse still, where our place was concerned, the evergreen shrubs that line the walk to the front door and give us protection from the road put out a second round of new growth after the inevitable fall pruning. No matter how hard those bushes are cut back each spring, there is always considerable cutting back to be done on our return. Usually I manage to make this duty into a long-term pleasure by giving the cut wands of juniper, yew, leucothe, and euonymous the glycerine treatment. This consists of mixing up a batch of pure glycerine (two tablespoons to every gallon of water) in a bucket and then smashing the last two inches of the stems of the clippings with a hammer and putting them in the bucket so they can draw the mixture up into their foliage. The combination of water and glycerine turns evergreen leaves an attractive bronze color and keeps the needles securely on the yews and junipers. Just as long as the vases in which these prunings are displayed are kept regularly topped up with the same mixture, I can use these prunings as part of the Christmas decorations.

Normally that second fall pruning is the end, for cold weather shuts off further growth. But that particular year many of the little dormant buds, which lie close under the bark and come to life in the spring to thicken up the pruned bushes, were fooled into premature fall growth. And, as with the new grass, not only did they not survive, leaving us with far fewer new shoots in the spring, but their very presence led to bad winterkill on the old branches. For the death of the out-of-season growth left open areas on the mature wood which were then also vulnerable to the cold. And this was the very last thing we wanted, since many of the large trees that also protect us from the street are in serious trouble because of the indiscriminate use of road salt during icy weather.

As always happens during an abnormally extended fall, there were one or two unexpected and extremely cold days. I was away when these fell upon the garden, and, most stupidly lured by the golden weather, I had left all the hydrangeas, house azaleas, and camellias outdoors to enjoy the extended summer.

None of the azaleas survived (and some were very old plants), and the camellias lost every bud. The only small comfort that I could glean from the wretched situation was that now at last I could give the camellias a desperately overdue reshaping without depriving myself of either buds or new growth.

The loss of the azaleas made for a very dull winter in the greenhouse, for it is never possible to buy any as large as old ones you have owned for years. But although I was annoyed with myself for being careless enough to lose those specimen plants, I had no sentimental attachment to them. All that was needed was time to grow new varieties to a more interesting size. The loss of the hydrangeas was a different matter. There were a good many of these. Some were florists' plants that had gradually been acquired over innumerable Easters, but the bulk of the collection was made up of the descendants of a single cutting of a blue lace-cap variety that I had filched, with permission, from a plant lent me for a daughter's wedding. The original parent of this plant had come from a private collection now, alas, dispersed. Since that particular friend is no longer active in gardening, I was extremely unhappy at the thought that through sheer carelessness I had deprived myself of a plant with so many associations.

Hydrangeas, however, are supposed to remain outdoors until

light frost kills the leaves, and although that vicious cold had done much more damage than just killing leaves, it seemed possible that some roots might have escaped, since hydrangeas are tougher plants than we realize. So instead of throwing them out, which was the fate of the azaleas, I put the wretched-looking hydrangeas into the cellar and treated them all winter long as though they were going through a normal dormancy—that is, I did not overwater, but neither did I allow the pot soil to dry out completely.

Nothing much came of all this effort until the end of March, when one out of a dozen pots put out a single weak shoot. This I coaxed into sturdier growth and eventually rooted as a cutting under lights. With much loving attention from me, by fall it had turned into a nice bushy plant. But I had no idea whether all this effort had been worthwhile, for there was no clue as to whether the saved plant was a plain old mophead or one of those cherished lace-caps. What's more, I did not expect to find out until the subsequent spring, for hydrangeas set bud on the wood of the previous season; they do not flower on new growth. But as sometimes happens, nature intervened and put me out of my misery. Late in September, possibly because the wood was by then sufficiently mature, a bud did form on a lower branch—lo and behold, it was a lace-cap!

The moral of this story is fairly obvious. No matter what the weather may be doing, in fall it is better to garden by the calendar than by the sun. If your important shrubs insist on growing, leave them alone if possible and make the spring cutting back even more severe. And take your house plants indoors at the usual period, even if a heat wave is on hand. For just as night must follow day, bitter cold is only just around the corner during a New England fall; it is foolish ever to take a chance.

A Parrot

In the recent long, extremely mild fall described earlier, the continued warmth and sunshine confused many plants, and they failed to slow down their growing pattern and get ready for winter. Among the plants that got into this muddle were the ivies that are used as ground-cover plants in the sheltered patio of the suburban garden. The continuous mild weather led these into a whole manner of excesses which included their trying to scramble up the trunks of the crab apples around which they are planted, where they are supposed to make a solid flat ring on the ground.

I do not like tree-climbing ivy, and I went to work pulling off these adventurous strands before they developed the heavy growth of suckers by which ivy glues itself to its host. If climbing ivy is tackled early it is simple to pull off. If, however, it once gets established, a saw and extremely strong clippers are needed to dislodge it. We already have enough problems with adventuring ivy in the country garden, where it is trying to take over a shingled shed as well as reestablish itself on an old elm. Keeping it under control has been an unending nuisance, for ivy dirties up the house and harbors all kinds of pests, and the shoots have enough hydraulic strength to force their way through and under almost anything.

I also happen to dislike the shaggy look of any tree trunk up which ivy is making its way, and nothing could be less appropriate than a tree covered with ivy in the patio area of the suburban yard where the entire visual impact is designed around formality and neatness. Ivy as far north as we garden also always brings with it an element of Russian roulette, for it is only *just* leaf hardy in our cold winters. Any prolonged spell of abnormally cold weather kills the so-called evergreen leaves. This means that when everything else is fresh and green with new spring growth, ivy can be covered with nothing but a mass of brown dead leaves. For although the stems themselves are rarely killed, fresh growth is extremely slow to break, and the dead leaves cling on tenaciously. Ivy on the ground can be partially protected from the hazard of intense cold by a light covering of leaves or by evergreen boughs, but tree-climbing ivy is impossible to protect, and more than once the appearance of our country garden has been spoiled right up to midsummer by the miserable appearance of dead ivy still clinging to the unreachable top of the shed or the high branches of the elm.

With all these problems I had no compunction whatsoever about nipping in the bud any ambition the ground-cover ivies might be harboring about getting up in the world, but behind all these practical reasons there is a more personal explanation for my antipathy to trees swathed in glossy ivy.

For many years my family owned a small green parakeet which as a little girl about nine years old I had trapped in a birdcage in a London square. I had noticed the bewildered parrot being mobbed by starlings for several days, and I had taken out some birdseed that was used to feed our canaries and spread that out on the ground for the parrot. It was obvious that the bird was starving, for it made ceaseless attempts to come down for the seed, but to my distress, every time it came slowly in for a landing, the starlings dive-bombed it and back it flew into the trees.

This enraged me, and I decided that since the parrot recognized birdseed it must be accustomed to cages, which the starlings certainly were not, and possibly if I put the seed in a cage instead of on the ground, the parrot might go inside to feed and thus avoid its tormentors. And this is exactly what happened. I brought out an old cage, spread seed on the floor, set the cage on a bench, and waited. The moment it spotted the cage, the parrot swooped onto it and, clinging tightly with its claws, worked around it head downward until it found the door. Since the cage was for canaries the entrance was small, but that didn't bother the parrot; it squeezed itself inside and began to wolf down the seed. But the starlings were not giving up so easily, and although they didn't like the look of the cage, they flew around it screaming furiously and closing in with tighter and tighter turns. The parrot was uneasy about the starlings, and it kept looking anxiously over its shoulder as it fed. I was in two minds about trying to drive away the other birds without also frightening the parrot, but when a starling became sufficiently bold to perch on the cage, the parrot itself took action. It ceased to feed, put its head out of the entrance to the cage, and deliberately pulled the door shut.

Such an obvious invitation to ownership could not be ignored. I picked up the tiny cage and took it home, and that was when Tugee, as we named him, joined the large menagerie of family pets. When, a couple of years later, we moved to the country he, of course, went too. But though we had bought him a large, rather grand parrot cage it had been a considerable struggle to keep him confined once he had recovered from his frightening adventure. For he was a highly ingenious bird and understood how to pick almost any birdcage lock. During the first few years we owned him we had to keep his cage door closed with a padlock, for he soon discovered how to unravel a twist of wire.

After we moved, the padlock to his cage got lost, and as a result, he was often discovered at large in the room where his cage hung. In the end, of course, he got out of his cage when a window was open, and our first intimation that he was again on the loose was a wild scream of delight when he spotted my mother weeding her flower bed.

He stayed outdoors for several days but eventually came back of his own accord to his cage, which we had hung outside. With that my mother decided that this confinement was absurd, the bird knew what he was doing and should be allowed his freedom.

From that time onward, his cage with the door open was hung in the conservatory, where a crack of a window was also always left open, and Tugee came and went at his pleasure. And for years he lived a lovely carefree life, sleeping in his cage at night if the weather was bad or cold but enjoying the freedom of the wooded Surrey countryside by day.

Then one warm November day, just as happened here in our last mild fall, the weather changed abruptly while he was outside; the temperature plummeted and a thick fog rolled in. Most unfortunately, the Christmas show of chrysanthemums had just been staged in the conservatory, and a new garden boy, who did not understand about the crack of window being left open for the parrot, shut it to preserve the show plants from the dirty fog.

Whether the fog and cold confused the parrot or whether, like Peter Pan, he tried to get in and found the window shut we never knew, but when he had not appeared by dark, even though my mother had reopened the window well before the light went, we became concerned and went outside with lanterns to try to coax him home. I even climbed an ivy-covered tree near the house to see if Tugee was perched on a branch where we often saw him spending the night outdoors in the summer.

But it was all no use. He was not to be found, and we never saw him alive again, although we did eventually find his body. For the following spring when tree men were working on that same ivy-colored elm, they discovered a little ball of very recognizable green feathers in a deep cavity that was entirely hidden from view by festoons of ivy.

It may have been the cold that killed him, or he may have died of old age. Whatever happened, it was a peaceful death in his sleep. But it made me very unhappy at the time to think that I had failed to rescue my old friend because of the thick ivy on that tree, and I suspect that part of my distaste for sprays of ivy draped over branches stems from that fact.

December

Hollies

In the country garden there exists a very dependable old tree of the local variety of holly (*Ilex opaca*) which, judging from its huge girth, must have lived through many changing seasons and survived several raging hurricanes and tidal waves. The tree perches rather precariously alongside the sea wall, where its roots are often submerged by abnormal winter tides. There are deep grooves all up the main trunk where, during the years when this piece of land was neglected, a hangman's noose of honeysuckle once twisted tighter and tighter. But it survived this hazard, from which we freed it with enormous labor, just as last year it survived a new and unexpected man-made danger when, unknown to us, the town saw fit to dig across its root spread in the course of laying a new storm sewer. And since this species is said to be only precariously hardy as far north as our yard, the tenacity of the old holly is all the more admirable.

At the proper season for as long as I have known it, now nearly thirty years, the tree adorns itself with gleaming berries.

When it was obscured by the jungle that sprang up following a hurricane, the berries were safe from all but foraging birds. But now that it stands free we have to mount guard over it at the holiday season, for the show it puts on is so lavish that the loaded branches can be spotted from the road.

Each spring when the old leaves fall I worry a little until the new growth comes in thick and bushy, but in general the old tree seems in excellent health and probably more than ready to outlast me. Since we cleared the land around it, giving it much more light, the new growth has unfortunately reached up into the utility lines that run overhead, and unless I am around to make a fuss, our fine symmetrical tree sometimes gets ruthlessly hacked back.

I resent overhead utility lines for their vulnerability in storms and for the way they disfigure the landscape. In time I trust that public opinion will become strong enough to induce the utility companies to lay these vital links underground everywhere, but until that moment comes, I do wish we could be given more consideration concerning the pruning that is deemed necessary to give the lines free passageway. Pruning is an art which should be learned by those who service the power and telephone lines to a much more sensitive degree. I accept the necessity of keeping the utility lines clear of overhanging branches and other such problems, but surely it cannot always be necessary to solve the matter by ruthlessly hacking a square opening in the canopy of a valuable tree—a practice now common everywhere.

For years, in spite of all those hopeful bright berries, the holly was barren; never a seedling did we find. Hollies are for the most part dioecious—that is, they are either male or female trees—and they cannot be told apart when young. That is why if you are buying very small hollies it is always wise to plant several, and it is better still to wait to buy hollies until they are old enough to have fruited so as to be sure of what you are getting. There are hollies, of which ours was an example, that will berry even without a male tree within wind-pollination distance, but in such cases the berries are often infertile, which again seems to have been the case with ours.

This situation changed dramatically when a neighbor set out a small holly bush—one of the newer hybrids. Its leaves were both darker and shinier than ours, and it grew in a more open form. It also flowered at an unexpectedly early age and, to the disappointment of its proud owner, turned out to be a male plant. But her loss was our gain, for once the new little holly bush started to assert its masculinity, we suddenly discovered innumerable seedling hollies in the rough ground near our old tree, showing that its long barren period was over. We were, of course, delighted with their appearance. However, they brought with them some new problems, for the place where they had germinated was near the high mark of those occasional freak tides, and although their old mother could, it seems, survive an occasional dose of salt water at the roots, it seemed most unlikely that the seedling trees could take such treatment. So since I did not want to lose these late-in-life offspring of our ancient holly—and as I also felt my neighbor should have some in gratitude for the virility of her little bush—we decided to move the most vulnerable of the seedlings into our growing-on area. Hollies move extremely badly, and the job is best done in the spring after some careful preparatory fall root pruning. But we did not have the time for the luxury of these options, for we didn't discover the seedling trees until the fall high tides were almost on us. Instead we dug the tiny sprigs with such enormous root balls that it seemed impossible that the roots could even sense that they had been disturbed, and they were lowered into predug holes in their new position as delicately as diamonds are set into a ring. But they sensed it all right; in the past three years, since Operation Holly Removal was initiated, they have put on hardly any new growth, and two or three have died. Since then we have discovered still more growing farther from the sea, and we have decided to leave these alone until they have reached a size that can take shifting better.

But, annoyingly enough, the place where this second batch has appeared has now produced another complicating factor that may do them in too. Minute holly seedlings need good dappled light to make strong growth, and they can be easily killed if rank growth smothers them. Ever since we have owned that particular stretch of waterfront land we have battled the encroachment of green ground elder. Down by the waterfront the high tides take care of the problem, and to prevent the pest from spreading past the area where we found it already established on the higher ground, we have kept a wide strip of land closely mowed as a path to act as a cordon sanitaire. Until recently this rather simple solution appeared to be entirely successful, but last year, to our dismay, the ground elder suddenly surfaced beyond the mown path into hitherto uncontaminated land. And where do you suppose those nasty little sprouts first showed up—among the seedling hollies!

So now we have to visit the area regularly and try to extract every scrap of newly sprouted root that we can dig up without disturbing those sensitive holly roots. If we can get the seedling trees past the first three or four slow growing years until they become big enough to rise above the ground elder all will be well, but it is going to be a long boring responsibility that I could have done without. And it will bring another problem in its train, for when and if the little hollies become large enough to be moved, what am I going to do about the pieces of ground elder that their root balls will inevitably contain? And will my friends feel that the acquisition of hardy, hybrid, berrying hollies, which with any luck I shall be able soon to offer around in quantity, are worth the hazard of introducing ground elder into their yards?

Waiting

A year or two back, we went down to the country garden in mid-December in the usual way to cut the holiday greens we use to decorate the town house. It had been an unusually mild long-drawn-out fall, and many of the deciduous trees and shrubs had held onto their leaves very late. But finally they were bare, for when we arrived the hard cold of winter was at last moving in and the ordeal of deep snow about to begin.

I always enjoy the leisurely job of walking about the grounds cutting branches from evergreens and red-berried shrubs that I have raised myself; it's an action that gives particular satisfaction. I also find this a very important moment visually in the yard. For without the distraction of flowers, leaves, or snow, it is possible to look at everything with a clear eye and evaluate the success of the basic planting plan. This is by far the best moment to decide whether the evergreens are positioned correctly for the most dramatic winter effect, and to see if the bare silhouettes of the trees could be heightened by selective pruning. So, as I walk around I carry strands of red yarn to tie on branches and bushes that I feel could do with a second look when spring leafs out. I take my time about the whole proceedings and in a relaxed comfortable mood make a point of visiting every part of the yard.

But that particular day I was not able to relax, for the familiar garden made a curious impression on me. By some odd combination of awareness and heightened sensitivity I sensed tension and expectancy everywhere as all the growing things in the natural world—trees, shrubs, plants, grasses—braced themselves like animals against the threat of the oncoming cold. The sensation came partly from the dark sky that hung very low overhead and the falling temperature and rising wind that chilled me as I walked around. In the woods I was struck by the almost unnatural brilliance of the ground mosses, their bright green tones suddenly reminded me of army banners lined up defiantly against an enemy. The yews hardly stirred in the rising wind; they rose tall and dark and almost menacing as they waited, like sentinels, to signal the coming storm. Down near the waterfront the big gray junipers were also on a winter alert. Their tremendous branches berried in blue usually provide safe chirping cover for innumerable small birds. But so electric was the atmosphere that the birds, if they were there, neither moved nor made a sound, and the only noise came from the wind menacing in off the sea, driving withered leaves through the dead grass. The feeling of danger, stress, and nervous energy loose in the garden was so vivid that I felt too restless and taut-nerved to go on cutting greens, and I turned back, almost with a sense of relief, empty-handed to the lights and warmth of the house.

During the night the enemy the garden was expecting attacked, and deep snow fell. When I went out in the morning the atmosphere had changed, the waiting was over, the sense of urgency had gone, and the garden had surrendered to a snow-enforced winter rest.

Probably I shall never experience again such a close personal association with the emotions that animate all living things when change or danger threatens—something we are now beginning to realize affects plants particularly strongly. Certainly I have never been tuned into it so vividly before or since—it was a memorable experience that I shared with our garden that day.

For Those with Eyes to See

Last Christmas the weather was very mild, and when we went to the country we found most unusual color for that particular season. The lawns were already glowing with the bright olive green sheen that is characteristic of the winter dormancy of cut grass in that area. The rough grass of the meadow had turned brown, but not its normal dingy shade—instead a vibrant brown that matched the leaves still clinging to the swamp oaks, so our small woodlot was circled with a bracelet of glowing copper.

During the leafless months evergreens gain enormously in importance as focal points, and those in the woods were looking particularly fine, the various shadings of their colors contrasting beautifully with the brilliant brown of the leaves and the grass. The white pines, having shed those sad pale needles that disfigure them in the fall, were their usual proud dark green. The steel blue junipers in the clearings were heavily berried, the small-leafed ilex was bright and shiny, and a big clump of rhododendrons looked particularly attractive. I was surprised to see how large this had grown, for in the summer I lose sight of this particular planting; it vanishes when the viburnums leaf out, and in the winter I am so used to seeing the bushes with the leaves curled in misery from the cold that I have ceased to think of them as a pleasant feature of the winter garden. Indeed I had almost forgotten, until I saw them that day, how handsome rhododendrons can look under high leafless trees when the winter weather suits them. These varied clumps, all of which I had laboriously set out over the years, had suddenly come into their own and were doing more than I could have expected of them to give form and texture to an otherwise empty landscape, and their strong colors were reflected in the dark ivy and purple ajuga running on the ground below them.

When the deciduous trees leaf out, we tend to forget the evergreens that do so much for the yard during the empty months. In our garden they are lucky if they get any food in addition to the natural nourishment provided by the decaying leaves that fall among them. But since a lot of the year is an empty period in this climate and one in which evergreens come into their own, they should not be neglected in this way but should be fed regularly and kept watered during dry spells. It is particularly important to be certain that young and shallow-rooted evergreens like rhododendrons do not go into the winter with dry soil around their roots. Plants and trees that hold leaves all winter never go entirely dormant but continue to transpire a limited amount of water. When the ground is deeply frozen, young evergreens find it impossible to replace the water given off by the leaves, since the roots that have not yet penetrated deeply cannot pull any in from the iron-hard ground. For that reason it is essential for their health that additional water be given before the ground freezes if the fall has been dry. Evergreens, for all their sturdy independence, are not invulnerable for many years after they are set out, and they need care as much as any houseplant.

But evergreens were not the only strong color contrast to the bare trees on that particular day, even though they were impressive. Deep in the woods there also exist innumerable enormous rocks, relics of a period when the area was under the sea. We can spot them only in the winter, for in the summer they are smothered under engulfing poison ivy. But when they gleam out gray, wet, and dark like huge stranded whales among the tangle of the leafless vines that net across them, they add an extra dimension to the winter scene. And their color is intensified by the contrast it makes with the brilliant flashing of the ground mosses that thrive in the swampy ground beside the paths.

But the most interesting additional color came from a source that hitherto I had never thought of in that connection. Scattered through the woods, on the rocks and on the trees, are innumerable lichens, which stood out with iridescent vividness. Some were fan-shaped in a delicate grayish green, some yellow deepening to rusty red, and others ghostly white, and by late afternoon everything was gleaming in soft sunlight.

Midwinter sun is usually either watery and weak or a threatening gleam that foretells a coming storm or harshly bright on fallen snow. But the sun that highlighted all the muted colors of the woods that day was sun of a different quality, soft and warm and more akin to summer sun. It was a lovely scene, a combination of chance factors that occurs but rarely, and I am glad that I happened to be there to see it.

Associations

During the long, enforced separation of World War II my mother used to make tissue paper roses that she mailed out to us each year across torpedo-infested seas. By some lucky chance none were ever lost, and we made them an important feature of our Christmas tree. The childrens' first real consciousness of their English grandparents came from opening the rather battered boxes in which each collection arrived and hanging the flowers, "from Granny and Grandpa in England," carefully on the tree. My mother was skillful with her hands, and she took an immense amount of trouble making those roses for the grandchildren she had yet to see. She tinted them carefully with watercolor paint so that they ranged from white through various shades of pink down to the deep red rose of England, and the petals were bound together with yellow thread to imitate the golden stamens.

We have some still, now rather tattered because they were made from the very poor grade of paper then available in wartime Britain, and they are still in use. But now it is my grandchildren who recognize them with delight as the Christmas ornaments are unpacked and hang them carefully on the tree just as their mothers did. I watch with great delight these simple flowers giving such long-term pleasure to three generations, for it was not only the children who loved getting those roses; they meant a great deal to me too.

But those fragile relics are only part of the memories that stir at this season, for we also possess the very first Christmas tree that we were all able to share. We moved into our present house the year the family was completed with the arrival of twin daughters. To celebrate that triumph and to mark the much longed-for acquisition of a house of our own, we decided to set out a living Christmas tree, at that period a rather unusual decision.

Winter was late arriving that year, and I can recall standing in the yard, with the snugly wrapped babies deep down in their double carriage, watching my husband, with the helping, hindering assistance of our small eldest daughter, dig the big hole into which the tree was to go after the festivities were over. It was a small spruce, and for many subsequent years it sulked rather uninterestingly beside the driveway, largely ignored except by me at Christmas when, unless deep snow hindered my passage, I usually made a sentimental pilgrimage to it. But then it took off, and now it is a magnificent specimen towering over everything else along that hedge. And this year, for the first time ever, it is carrying a heavy crop of cones showing that it has reached maturity.

Today the family is scattered, and some of them celebrate Christmas far away, establishing their own family traditions in totally different settings. But whether the house is filled with grandchildren or whether Christmas is spent alone, we still have the paper roses and that splendid tree, symbols of the past, present, and future and the touchstones of countless warm, happy memories.

Now it is more important than ever to plant trees, not just to mark special occasions but as a part of our concern for the landscape and for the environment. Recently I read to my delight an account of how a community banded together to save, through an elaborate moving operation, a 175-year old ginkgo tree that had the misfortune to be growing on the route of a new road.

Not all of us have the opportunity to participate in quite such a dramatic display of our concern, but we can all do our share toward the conservation of fine trees everywhere through joining societies involved with their preservation and adding to the number of trees in our own community by planting a living Christmas tree if we are blessed with a yard. One sacrifice that will have to be made as a result of this decision is that the tree must be small —something that can be a positive blessing where expensive ornaments are involved. And particular care of it must be taken during the time that it is indoors. You will also have to forgo the rather dubious pleasure of keeping a tree indoors for a long time after the holiday is over. But with these few small changes in style you can have your cake and eat it too, just as we did years ago.

Christmas trees with a root ball encased in burlap can be bought in December at almost all good garden centers. The various spruces take to indoor life the best, but no living tree should be kept in the house for more than a week at the most. This means that to get a good choice an early-bought tree must stand somewhere outdoors, but since a soggy snow-covered root ball is most unpleasant to bring indoors, the tree should stand under cover in an unheated place. It cannot be very tall, for a root ball that is manageable indoors will support a tree only about five feet tall. The effect when you have it in use raised up on a strong table is not unlike the mid-nineteenth-century pictures of Christmas trees in children's books. Indeed, the Christmas trees I remember from my twentieth-century childhood were still always smallish trees that stood on tables. Cutting huge evergreens before the days of Christmas tree farms was considered a shocking waste of something that in nature develops rather slowly.

While the tree is indoors it is vitally important that it stand in a solid container so that water can be applied to the root ball; an old roasting pan prinked up with foil will do excellently. The tree should also be kept sprayed with lukewarm water as much as is compatible with not ruining the decorations, and it is important to arrange the lights so that they do not burn the tender foliage tips. But there is another aspect of having an indoor living Christmas tree that is even more vital than the treatment it is given in the house, and that is to have the hole where it is to be planted dug well ahead of frost and the soil from the hole stored in a covered container in a frost-free place. It is also a wise precaution to store dampened peat moss in the same place in yet another container, together with a couple of armfuls of wet leaves.

Since it is impossible to judge exactly how large the root ball of a tree may be, and the hole should be dug in the north by November, make the hole far larger than seems likely to be needed and take a rough measurement of its overall diameter before you go shopping for the tree, for if the ground has frozen hard you will not be able to enlarge that hole. If you live where there is a likelihood of snow before Christmas, cover the excavation with plastic held down on all four sides with stones—don't use earth, for that may freeze over.

After Christmas, and just before planting, mix up a bucketful of the stored soil and damp peat moss and dump it into the bottom of the hole. If possible, numb hands and all, do the whole planting operation in one day; you never know at that season whether a sudden change of weather may not prevent your finishing a half-done job. Lower the burlapped root ball into the hole and test whether the knot reaches almost to the soil surface. If it does not, the tree will have to come out and more of the stored soil be put in underneath. Since Christmas is a family festival, there will hopefully be extra hands to help you lift the tree around. If it is dragged there by the trunk, and levered in and out of the hole again by its neck, untold damage will be done.

With the tree in place, pour the stored soil between the sides of the root ball and the edge of the hole and thump this down with the handle of a rake. Then cut the ropes and turn back the burlap. If the hole is way too large for the root ball, extra soil may have to be shoveled in and compacted by thorough trampling. Cover the top of the root ball with about an inch of soil, then throw an inch or two of mulch made from those stored wet leaves over the entire planting area.

After that there is nothing to do but wait. Hopefully the tree will not die, but it will certainly relapse into shock, coming from a warm house and going into cold ground, and you should not expect anything spectacular in the way of growth for several years. There may, in fact, even be dieback the following spring, but as long as some new growth appears all will be well. If when the ground thaws the tree sinks a little, pull back the covering mulch so that the roots are less deeply buried. In general, however, a thick mulch over the planting hole will do a great deal for the tree by retaining moisture, and the mulch should be kept regularly renewed for several years.

This is a small, positive contribution we all can make to counter the disposition that has taken place in our forests. Most of us don't have enough space to make this Christmas tree planting an annual operation, but we should think about doing it for some special occasion and to give us very special lasting pleasure.

Christmas

The links that bind us to our past are perhaps cherished more at this particular holiday season than at any other time. Some of us try to re-create what we remember while others try to establish new traditions. There is something to be said for both approaches, for they serve as a cohesive force within a family; they imprint themselves particularly upon the minds of our children, and our actions at this season become part of their heritage.

As a child I had very little home exposure to conventional Christmas decorations. My family disliked tinsel and artificial ornaments, as I still do, and my mother was strongly opposed to the use of Christmas trees indoors. This was not because she had any puritanical objections to Christmas gaieties but merely because she considered the indoor trees extremely dangerous. At that period they were lighted with small wax tapers in flimsy tin holders clipped to the branches, and every family had some horror story connected with the fires caused by these tiny open flames. I have vivid memories of the fierce precautionary warnings I always received before going to parties at other people's houses where there might be a lighted tree. The orders were to stand beside the door and well away from the tree the moment the candles were lighted. Part of the danger came, of course, from the dried-out needles of the tree itself, and part from the fact that little girls in those days wore wide-skirted flimsy dresses. At children's parties it was also the custom to hang toys from the tree which were given out as favors to the guests, and children pressing forward to get their presents sometimes inadvertently came into contact with the dangerous flames.

To this day a Christmas tree means far less to me than the other holiday symbols, and enough of my mother's attitude still lingers so that ours comes in as late as possible and is banished long before the traditional Twelfth Night. I also insist that it stand in a well-filled bucket of water all the time it is indoors to minimize drying out.

Cutting the greens from our own backyard is far more important to me, for this action takes me back to my childhood, when long strands of ivy and bright shiny-leafed red-berried holly from the hedges were brought indoors together with a few sprigs of mistletoe from a very old apple tree at the end of the garden. These greens were arranged in a huge bouquet in the middle of the mantelpiece in the room where the presents were opened, and for an extra festive note, sprigs of holly were also tucked behind all the picture frames and mistletoe hung over the doorway. The only other special decorations were in the dining room. Here, in the center of the table, there was usually a bowl of Christmas roses if the season for these flowers had been a good one. If there were none, my mother bought bunches of bright red St. Brigid anemones from the florist for the centerpiece. Around this bowl we laid another garland of home-gathered greens—this time holly, ivy, and yew, for mistletoe was considered unlucky on the table. Legends cluster around mistletoe, for it has been held either sacred or magical from a very early period, and even my highly practical parents kept to the old ways where mistletoe was involved.

Fixing the garland in the dining room was always something of a strain, for it had to be arranged extremely carefully on a special tablecloth that appeared only at Christmas. No staining with the berries or bringing in of clinging dirt on the ivy was allowed. The cloth had been embroidered by my great-grandmother with the same Christmas symbols of holly, ivy, and mistletoe and is extremely fragile. I use the present tense, for I have it still. It survived a wartime sea journey when my mother sent it over as a special present. We still use it at Christmas, but since I grow no Christmas roses we have a centerpiece of fir cones, dried nuts, weed pods, and other natural material picked up in our yard and built into an elegant permanent decoration by one of our daughters. Around this we also spread a wreath of home-picked holly and ivy and yew—with all the same fidgety warnings from me. The continuity this tradition represents is something I cherish.

We have a second family custom for this season which came about by chance. One Christmas I bought some chunks of lightweight feather rock to form the cave for our crèche figures. The following spring I had lots of small divisions of African violets going to waste, and since I had read somewhere that these plants do well on pumice rock, I decided to experiment. With care, for pumice is as sharp as glass, I chiseled out a number of deep holes, filled them halfway up with rich compost, and set in the little plants. With good light and an occasional dose of weak liquid fertilizer they flourished amazingly; the leaves spread across the rock, and there were literally hundreds of buds. By better luck than skill the bloom was at peak the next Christmas, so that year instead of a plain rock shelter the crèche was housed in a flowery bower that could have come out of a Botticelli landscape. The entrance was overhung with foliage, flocks of wooden sheep grazed happily among the white and purple flowers, while kings and camels strode over the leaves. The following year the rocks were again plain, but they were greeted with such cries of disapproval that I now always try to plan bedecked feather rocks for Christmas. These are now firmly associated in the minds of the grandchildren as part of the Christmas scene.

With a little forethought even the most housebound gardener can think up something to be done with plants and flowers to mark one of the great seasons in our calendar, and it is an easy way to establish a tradition that will involve us with the natural world.

Waste Not

Whenever I pass trash barrels awash with discarded Christmas trees I feel depressed, for this is yet another example of the casual way we waste living things and add to the litter problem that is engulfing us. Christmas trees these days are cultivated as a moneymaking crop, so the huge trucks, with trees stacked up like matches, that rumble across country to city markets for weeks before the holidays are no longer evidence of the ruthless spoilage of our woods. But cultivated or wild Christmas trees still cost plenty, and most of us do not get as much use out of them as we should. Considerable lasting enjoyment is available even from a cut tree, if some precautions are taken when it is bought and after it is taken home.

The most important point is to make sure that the tree you buy is reasonably fresh, for the harvesting of trees starts long before Thanksgiving in some areas. One choice which is becoming increasingly popular is to find a Christmas tree farm where you can cut your own tree, something children love to be a part of. The alternative is to be extremely fussy about the tree you buy, with particular emphasis on two matters that have nothing to do with the size or the bushiness of your choice. With reasonably fresh trees there should be moisture still oozing from the cut stump, and the needles should not shed abnormally, even when you test this by banging the tree hard up and down on the pavement. Obviously every tree will shed a few needles with this rough treatment, but if they fall to the ground in a shower, reject the tree no matter how elegant everything else may look. Heavy needle shed will not only get far worse when you bring the tree indoors, driving you mad with the mess, but it is also a sign that the tree is already drying out, which can turn it into a dangerous fire hazard in the dry atmosphere of the house.

Once you have found a tree that has a fresh-cut stump and is holding its needles, try to store it either outdoors or somewhere that is very cold until the moment comes to bring it indoors. You can also prepare it better for the difficulties ahead by sawing an additional section off the base and standing the tree with the freshly cut stump in a bucket of water.

If you have no cool place where the tree can wait, it is better to put off buying it until the last possible moment. This may mean that you will not get such a good choice, but the tree you do bring home will still be fresher from having stood outside at the retailer's than a more elegant specimen that has waited in a heated hallway. Indoors, for safety's sake, the tree should always stand in a container that holds water, and even a tree bought at the last minute should always have the base recut. In all conifers a resinous sap oozes out of the bark after any injury—this, by the way, is the so-called moisture that you should look for as a sign of freshness. But in an attempt to conserve the moisture that remains in the tree, this sap soon hardens into a watertight seal which will prevent the tree from taking in more water even though the cut end is submerged in the bucket.

With the base recut the seal is broken, and a Christmas tree will take in an astonishing amount of water, particularly if the container is filled with lukewarm, rather than icy cold, water. Moisture being drawn up into the tree is what gives the room that lovely forest smell; the additional water also helps the tree retain its needles and minimizes the danger of a flash fire from resinous dried-out needles. But don't forget to check the container regularly while the tree is indoors, for in our hot houses a tree will use up extra water at considerable speed.

I don't happen to enjoy seeing a tree indoors after the Christmas festivities are over; I think they look messy, and I am always nervous about fire. But above all I like to get extra use out of our tree, and the longer it stays inside the shorter the period in which it will continue to look well outdoors afterward.

There are several ways of reusing an old Christmas tree. It can be "planted" in a large plastic pot, held steady with stones and sand, and used to decorate a balcony or doorstep for many months. Cut trees, which sometimes can be had for the asking after Christmas, can be used this way to serve as an attractive matched pair of evergreens. I have had them remain in excellent condition with their cut ends stuffed into earth in plastic pots until the weather warms up in late March.

Old Christmas trees can also be used in the yard tied to a fence or a bare-leafed bush to provide shelter for birds. Three, tied wigwam fashion, near a feeder will enormously increase the number of birds you attract. In modern airport terminology they provide a holding pattern for birds low in the pecking order, as well as a shelter from bitter wind and a place where snow does not cover the ground so there is room to provide food for ground feeders.

For horticulturists cut branches of Christmas trees make an excellent mulch for bulb plantings that may have got a little ahead of themselves or for ivy ground covers in cold climates.

But with all this advice about reusing trees, may I tack on one urgent request? Whatever you do, please pick off all the derelict tinsel. Offhand, I can think of no useful function this can serve once the decorations are put away, and it looks perfectly dreadful garnishing a yard!

Enchantment

One year, after a long, unusually mild season, we went down to the country garden late in December during the first slight snow flurry of the season, which was followed during the night by a squall of freezing rain. And in the morning every twig and shrub was outlined in icy transparency and bowed down with the weight of frozen snow. It all looked entrancing, but some of the bushes near the house were dangerously spread open by the weight of the ice, their branches resting on the icy ground, where they would soon be frozen down.

Old bushes recover from this winter hazard without much damage except to the tip ends of some of the branches, but I was concerned about some shrubs which I was not anxious to have welded down by frost quite so early in the snow season. These were some big evergreens and azaleas that we had moved a couple of months previously, taking them from an overcrowded shrub border and planting them in a fresh curve of blossoming trees and shrubs farther from the house where they would have far more room to expand and could delight the eye better in late spring.

Moving them had been a considerable undertaking, for they were large, well-established plants. But since we had the good fortune to have gardening visitors aching for action staying in the guest house, and moving these shrubs was long overdue, we had decided to take a chance on the weather and make the shift even though the work took place a little later in the year than was ideal. But since the shrubs had been reset so recently it was important to make sure that the early ice and snow was not weighing them down, for newly moved bushes do not have the resiliency of well-established plants, and it seemed likely that a helping hand might be needed to tie up the foliage and prevent ice spread. This, of course, should have been done ahead of the snow but had unfortunately got overlooked. In general, it is unwise to try to interfere with ice or snow on trees and shrubs, for cold makes branches extremely brittle and heavy-handed shaking often does more harm than good. But this, however, was something of an emergency, so I gathered up some rope and went to take a look.

The grass was sharp as glass as I crossed over it, the air was quiet and still, and the sea beyond was gray and lifeless. It was the first touch of real winter, and cold had fallen over the garden like a shroud. The new curve into which we had set the shrubs lies well back from the sea behind a stand of swamp maples and white pines, in a part of the meadow that abuts on the woods. The area has almost the feeling of a small clearing in the forest, for although it is open land it is surrounded on three sides by big trees.

When I got there it could have been another world. By some quirk of the air currents there was no snow on the ground, and ice had not formed on anything. The grass was soft and green, dandelions flashed yellow, and the air was full of the call of sheltering birds. The recently moved bushes needed no shaking and no roping. Nothing was amiss.

Legends and fairy stories tell of men out walking who are

bewitched and find themselves suddenly in strange places of sound, light, and enchantment. I now feel that I know how those stories arose, for on that gray cold morning I too had the experience of passing from winter into what could have been an enchanted world.

Index

achimenes, 5
aerial sprays, 165–8
African violets, 4, 118, 295
air layering, 268–9
amaryllis, 262–7
Amelancher canadensis (shadblow or serviceberry), 68
annuals
  fall care, 213–14, 226
  seed, saving, 215–20
  seedlings, self-sown, 219
aphids, 115, 270
apple trees, 68, 261–2
  blossoms forced indoors, 60
Asclepias tuberosa (butterfly weed), 69–70
ash trees, 230
asparagus fern (sprengeri), 23
asters, wild, 174–5
azaleas, 46, 273
  house plants, forced, 3–4

Ballou's Dollar Monthly, 44
beans
  Indian method of growing, 184–9
  Kentucky Wonder, 186, 187
  scarlet runner, 188–9
bees, 182
Beeton, Mrs., *Household Management*, 129
begonias, 122
  tuberous, 5, 122, 193–200
biennials, 148–52
Bigelow, Albert S., 189
birches, 230, 233
birds, 36–8
bittersweet, 142, 231

black-eyed Susan, 173, 174
Boston daisy (marguerite), 183
bridal wreath, 260–1
bulb fiber, imported, 8–9
bulbs
  fall planting, 249–51
  forcing in cellar, 65–6
  narcissus, in house, 7–10, 251
  potted, in cold frame, 64–6, 251
  storage, 5, 6, 65, 251–2
butterfly weed, 69–70

cacti
  Christmas, 5, 113, 252
  ferns in pots with, 12–13
  orchid, 112–16
  Thanksgiving, 252
camellias, 273
Campanula medium calycanthema, 149, 150, 152
Canterbury bells, 149, 150, 152
Cape cowslip (lachenalia), 247–8
cats
  birds and, 37–8
  plants eaten by, 23
cedar, red, 230, 232
chicory, 173
children, gardening lessons for, 57–9
chionodoxa, 249
chipmunks, 212
chlorosis, 269–70
Christmas, 292–5
  crèche, 294–5
  greens for, 284, 293–4
Christmas cactus, 5, 113, 252
Christmas roses, 294
Christmas trees, 292–3
  cut, care of, 296–7
  discarded, 295–8
  living, planted, 289–92
  paper roses for, 288–9
chrysanthemums, 239–40
  cuttings, 236–7
  fall plantings, 215
  forced for out-of-season bloom, 234–5
  pinching off tips, 237
  potting for winter, 238–9
cinnamon fern, 111
clivias, 29–34
  berries, 31
  division, 30–1
  resting period, 5, 6, 32
Coates, Alice, *Flowers and Their History,* 174
cold frames, bulbs in, 64–6, 251
columneas, 12
compost, 71, 108–9, 256–60
conservation
  aerial sprays, 165–8
  fall foliage and, 242
  leaves for compost, 256–60
  planting trees, 290–2
  road construction and, 69–71
  salt marsh damaged, 153–5
  wild flower preservation, 131–2
conservatory, 52–4
corn
  Indian method of growing, 184–9
  raccoons eating, 210–11
cosmos, 183
crab apples, 68, 230, 233
  blossoms forced indoors, 60
cranberry bog, 172–3
crèche, 294–5
crocus, 66, 249

cutting
  chrysanthemum, 236–7
  geranium, 125–7
  ivy, 134–6
  water-rooted, transferring to soil, 14–16

daffodils, 232
  bulbs, planting, 249–50
  indoors, 251
  miniature, 66–7
daisies
  Boston (marguerite), 183
  oxeye, 143
Dana, Mrs. William Starr, *How to Know the Wild Flowers,* 129–30
day lilies, 225
  lemon (*Hemerocallis flava*), 143
  tawny (*Hemerocallis fulva*), 144–5
deer, 212
Dianthus barbatus auriculaeflorus (sweet william), 149
dodder, 166
dogtooth violets (trout lilies), 71–3
dogwood, 60
  kousa, 17–18, 225–6
Dracaena, 269

Easter lilies, 74–7
echeverias, 122
electric lines, overhead, 281
enclosure, glass-roofed, for plants, 52–4
epiphyllum (orchid cactus), 112–16
Erythronium americanum (trout lilies or dogtooth violets), 71–3
evergreen shrubs, 261, 272–3, 286–7
  glycerine treatment for, 272
  ice and snow on, 298–9
evergreen trees, 16–17, 284–7
  roadside plantings, 68–9

fall care of gardens, 213–15, 225–7, 248–53, 271–4
fall colors of foliage, 240–2
fatshedera, 141
ferns, 109–11
  cinnamon, 111
  as ground cover, 109–11
  as house plants, 122
  interrupted, 111
  maidenhair, 13, 122
  Osmunda, 111
  in pots with cacti, 12–13
  royal, 111
forests, 24–5
forsythia, 59–61
foxgloves, 149, 152
fruit trees, forcing flowers indoors, 60

Galanthus nivalis (snowdrops), 27
Galanthus plicatus (snowdrops), 27
gardenias, 122, 269–70
garden notebooks, 46–9
geese, wild, 231
gentian, fringed, 169
geraniums, 121–2
  cuttings, 125–7
  pinching tips, 126–7, 138–9

geranium (*continued*)
regal (Martha Washington), 123–8
repotting, 137–8
resting period, 5, 6, 137
winter, 136–9
gladiolus corms, storage, 5, 206
gloxinias, 202–9
forced for bloom, 4, 203
outdoors, 207–8
resting period, 5, 203–6
starting, 203
winter growing treatment, 204–5
glycerine treatment for evergreens, 272
goldenrod, 173–5
goutweed (ground elder), variegated, 225
grapes
pruning, 38–40
raccoons eating, 211
wild, 231
grass clippings, 107–9
greenhouses, 54–6
ground cover
ferns, 109–11
ivy, 78–80, 275–6
ground elder, 46, 85–6, 181, 233, 283; variegated, 225
groundhogs, 212
Guernsey lilies (nerines), 243–7

hawkweed, 143
heaths and heathers, 85–9
Hemerocallis flava (lemon day lily), 143
Hemerocallis fulva (tawny day lily), 143–4
herbicides, 70
Hippeastrum, 262–7
holly, 233, 280–3
for Christmas, 293–4
honeysuckle, Japanese, 142, 144–45, 231
hornets, 182
horse chestnut, 233
hosta, 111, 225
house plants
conservatory (glass-roofed enclosure), 52–4
forced for bloom, 3–4
outdoors, 118–20
resting period, 5–6
root pruning, 120–1
in summer, 118–23
vacation care, 161–3
in winter, 252
hurricane, regeneration after, 230–4
hyacinths
bulbs, planting, 249
indoors, 251
Roman, 251
hybrid plants, 219
hybrid seed, 216, 220
hydrangeas, 273–4
resting period, 5–6

Ilex opaca, 280
impatiens (busy Lizzie), 122
Indian method of growing vegetables, 184–9
insecticides, 36
aerial sprays, 165–8
interrupted fern, 111
Iris reticulata, 66
ivy, 77–80
for Christmas, 293–4

ivy (*continued*)
cuttings, 134–6
flowering, 140
as ground cover, 78–80, 275–6
as house plant, 134–6
petunias with, 132–4
tree-climbing, 275–6
tree or standard, 141

joe-pye weed, 173
juniper, 68, 230, 233, 285

Kartuz, Michael (expert on gesneriad family), 204
kousa dogwood, 17–18, 225–6

Lachenalia (Cape cowslip), 247–8
lawns
fall care, 249, 271–2
mowing, 190–2
watering, 192
leaves for compost, 256–60
Leighton, Anne, *Early American Gardens*, 107
leucojum (snowflake), 27, 28
lichens, 287
lilacs, forcing flowers indoors, 60
lilies
Easter, 74–7
Madonna, 77
lobelia seedlings, 219
loosestrife (*Lythrum salicaria*), 168–70

Madonna lilies, 77
maidenhair fern, 13, 122
maples
sugar, 242
swamp, 230
marigolds, African, 183
Martha Washington geranium, 123–8
May apples, 232
Mayflowers, 130, 132
mealybugs, 29, 266, 270
Michaelmas daisy, 174
mistletoe, 293–4
monstera (Swiss cheese plant), 21, 22
mosquito control, aerial sprays, 165–8
mulch, 162–3, 227, 237
on biennials, 151
grass clippings, 107–9
newspaper, 109
in tree planting, 292
in vegetable gardens, 186, 189

narcissus
paper-white (polyanthus), 7–10, 251
soleil d'or, 251
nerines (Guernsey lilies), 243–7
Nerine sarniensis, 243
nettles, 104–7, 240
newspaper as mulch, 109
notebooks, garden, 46–9

oaks, 230, 233
Oasis (flower arranging material), 135

opossums, 211–12
orchid cactus, 112–16
orchids, 13–14
  wild, 131
Osmunda ferns, 111
overhead utility lines, 281

parrot, 275–8
parsley, 101–4
peat moss, 163, 195–7
peat pots, 121–2
Pelargonium domesticum (regal or Martha Washington geranium), 123–8
perennials, fall care, 214, 226
petunias, 221
  with ivy, 132–4
  with parsley border, 101–4
philadelphus, 233
philodendrons, 20–4
phlox, 183
  seed not worth saving, 218–19
plant lice, 134–5
plastic bags, plants stored in, 161–2
plum, flowering, 89–90
poison ivy, 231, 240
  introduction in England attempted, 145
pond, 80–5
poppies, Shirley, 216
primroses, forced, 4
Primula malacoides seedlings, 219
privet, 233

Queen Anne's lace (wild carrot), 144

raccoons, 210–11
ragged robin, 144
red spider, 265
regal geranium, 123–8
rhododendrons, 286–7
roadsides, native plants and landscaping, 67–71
Roman hyacinth, 251
roses, 141–2
  rugosa, 233
royal fern, 111
Royal Horticultural Society, *Dictionary of Gardening*, 107
rudbeckia, 173, 174

salt, effect on plants, 45–6, 181
salt marsh, 153–5, 165–6
  regeneration after hurricane, 230–1
salvia, 222
scale insects, 134, 270
scarlet runner beans, 187–8
scents, memory and, 140–2
scillas, 249–50
seed
  home-grown, saving, 215–20
  hybrid, 216, 220
shadblow (serviceberry), 68
Shirley poppies, 216
skunks, 210
slugs, 114, 150
snowdrops, 25–9
  bulbs, planting, 28–9, 249
snowflake (leucojum), 27, 28
South African bulbs, 242–8, 262
squash, Indian method of growing, 184, 186, 187
squirrels, 197–8, 212

Sprengeri (asparagus fern), 23
sunflowers, cross-pollination, 217–18
sweet william, 149, 152
Swiss cheese plant (monstera), 21, 22

tamarack, 230
tansy, 144
television gardening programs, 49–51, 204, 268
Thanksgiving cactus, 252
toads, 34–6
tomatoes, planting seeds indoors, 57–9
trees
  bonsai training, 82
  fall colors, 240–2
  oxygen from, 257–8
  planting, 289–92
  in winter, 16–18, 284–7
trout lily (dogtooth violet), 71–3
tuberous begonias, 122, 193–200
  grandiflora, 195, 196, 199
  hanging types, 194–5
  multiflora, 195, 196, 199
  starting, 194–6
  tubers, storage, 5, 200
tulips, 98–101
  bulbs, planting, 250–1
tupelo, 230, 233

vacation care of plants, 161–3
vegetable gardens, 156, 163, 188–9, 222–3
  Indian method of growing, 184–9
  mulch, 186, 189
veronica, 182
Virginia creeper, 145, 241

Wallace, Henry A., 187
wasps, 166
water, 155–9
  cold, plants harmed by, 157–8
  rainwater, 159
watering, 157–9, 163, 192
water lilies, 82
wedding receptions, 89–91
weeds
  fall weeding, 226–7
  introduced from other countries, 145
White Flower Farm, 8
whitefly, 127–8
white pine, 230, 233, 286
wild flowers, 129–32
  American, cultivated in Europe, 173–4
  introduced from other countries, 143–5
  preservation, 131–2
  summer, 172–6
wildlife, 209–13
Wilks, Rev. W., 216
wisteria, 177–80
witch hazel, 60
woods, 24–5

yew, 285, 294

zinnias, 221

A Note About the Author

Thalassa Cruso was born in 1909 and spent most of her childhood in Guildford, Surrey. She was trained in archaeology and anthropology at the London School of Economics, where she took an honors diploma in 1931. After apprenticing under Sir Mortimer Wheeler at Verulamium (St. Albans) and Professor Christopher Hawkes at Colchester, she excavated and published a report on the Iron Age Fort at Bredon Hill in Worcestershire. From 1931 to 1935 she was an Assistant Keeper at the London Museum in charge of the Costume and Nineteenth-Century Collections and the author of a book on costume. During World War II she worked for the British Consulate in Boston, where she has lived since her marriage in 1935. Throughout her varied career she has maintained an active interest in horticulture. In the fall of 1967 she launched a very successful continuing television career with "Making Things Grow" on WGBH-TV (Boston) and five affiliated New England educational stations. She is a Fellow of the Society of Antiquaries of London, a member of the Royal Archaeological Institute, the Royal Horticultural Society, the Garden Club of America, the Garden Club Federation of Massachusetts, Inc., and the Massachusetts Horticultural Society, and is a horticultural judge and the winner of many gardening and greenhouse awards. In 1969 she was awarded the Garden Club of America's Medal of Merit by the Chestnut Hill Garden Club. In 1970 she was the recipient of the Horticultural Society of New York's citation for distinguished horticultural service and the Garden Club of America's Distinguished Service Medal.

A Note on the Type

The text of this book was set on the Linotype in Aster, a typeface designed by Francesco Simoncini (born 1912 in Bologna, Italy) for Ludwig and Mayer, the German type foundry. Starting out with the basic old-face letterforms that can be traced back to Francesco Griffo in 1495, Simoncini emphasized the diagonal stress by the simple device of extending diagonals to the full height of the letterforms and squaring off. By modifying the weights of the individual letters to combat this stress, he has produced a type of rare balance and vigor. Introduced in 1958, Aster has steadily grown in popularity wherever type is used.

This book was composed, printed, and bound by Kingsport Press Inc., Kingsport, Tennessee
Typography and binding design
by Carole Lowenstein